Engineering Drawing Problems Workbook

Series 4

to accompany

Technical Drawing with Engineering Graphics

Paige R. Davis
Louisiana State University

Karen R. Juneau
William Carey University

Prentice Hall

Boston   Columbus   Indianapolis   New York   San Francisco   Upper Saddle River

Amsterdam   Cape Town   Dubai   London   Madrid   Milan   Munich   Paris   Montreal   Toronto

Delhi   Mexico City   Sao Paulo   Sydney   Hong Kong   Seoul   Singapore   Taipei   Tokyo

CONSTRUCT EACH LETTER AND NUMBER THREE TIMES USING THE BOXES PROVIDED. LETTER
EACH OF THE NOTES ON THE GUIDELINES PROVIDED BELOW THEM. USE AN HB OR F LEAD.

A   B   C

D   E   F

G   H   I

J   K   L

M   N   O

P   Q   R

S   T   U

V   W   X

Y   Z   L

2   3   4

5   6   7

8   9   0

1020 STEEL FAO   FILLETS & ROUNDS R3 UOS

M42X4.5 HEX HD BOLT   REG LOCK WASHER

$6\frac{1}{10}$ $2\frac{5}{9}$ $3\frac{7}{8}$

NAME:

SECTION:

INTRODUCTION TO
MODERN GRAPHICS

PLATE
1

1. COMPLETE THE TITLE BLOCK SHOWN BELOW.
   ALL LETTERING IS 1/8" EXCEPT FOR THE 3/16"
   DRAWING TITLE. BE SURE TO INCLUDE YOUR
   NAME, SECTION NUMBER AND DATE IN THE
   APPROPRIATE LOCATIONS.

DRAWING TITLE	DR. BY:		SEC:
	SCALE:	DATE:	NO.

2. COMPLETE THE TITLE BLOCK BELOW.
   ALL LETTERING IS 1/8" EXCEPT FOR THE 3/16"
   DRAWING TITLE. BE SURE TO INCLUDE YOUR
   NAME, SECTION NUMBER AND DATE IN THE
   APPROPRIATE LOCATIONS.

LOUISIANA STATE UNIVERSITY BATON ROUGE, LA		
DRAWING TITLE		
NAME: YOUR NAME		SEC:
SCALE: FULL	DATE:	1 OF 4

3. COMPLETE THE PARTS LIST.

5	COTTER PIN	1	
4	FRAME	1	STEEL
3	WASHER	1	BRASS
2	PIN	1	STEEL
1	ROLLER	1	CI
NO.	PART NAME	REQ'D	MATERIAL

NAME:	INTRODUCTION TO MODERN GRAPHICS	PLATE
SECTION:		3

DRAW OR SKETCH EACH LINE STYLE IN THE SPACE BELOW THE EXAMPLE.

VISIBLE LINE
HB — .6mm

HIDDEN LINE
HB — .3mm

CENTER LINE
2H — .3mm

CUTTING PLANE OR
VIEWING PLANE
HB — .6mm

CUTTING PLANE
HB — .6mm

SHORT BREAK LINE
HB — .6mm

SECTION LINE
HB — .3mm

DIMENSION &
EXTENSION LINES
2H — .3mm

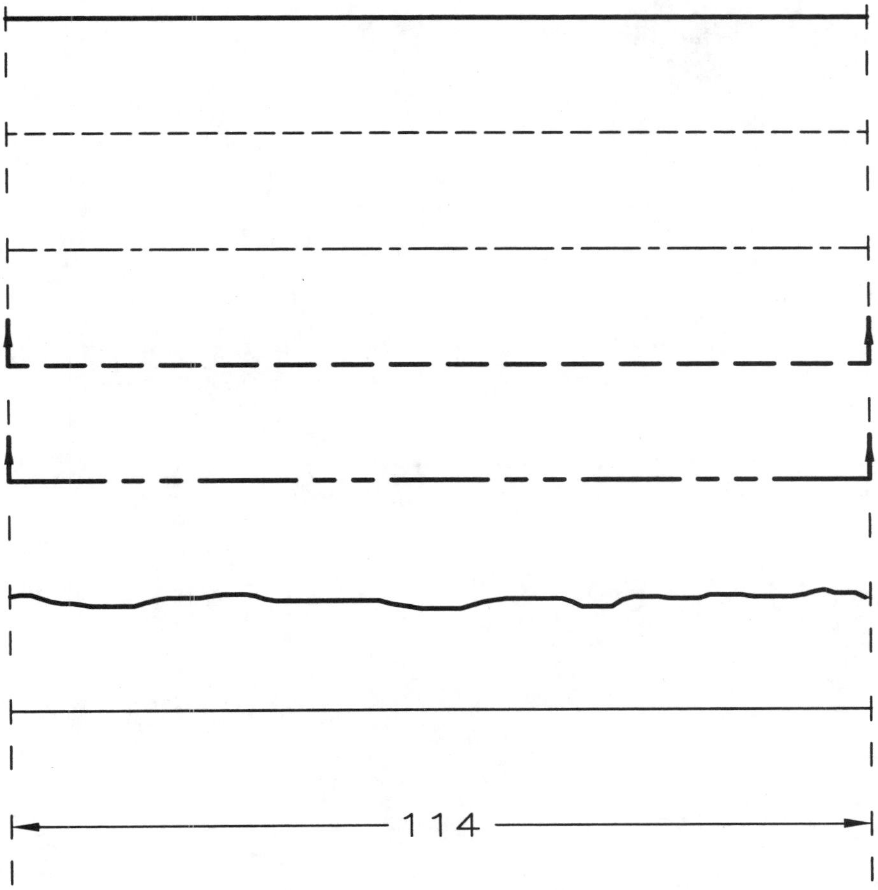

114

IDENTIFY EACH OF THE FOLLOWING LINE TYPES SHOWN IN THE FIGURE SHOWN BELOW.

A _____

B _____

C _____

D _____

E _____

F _____

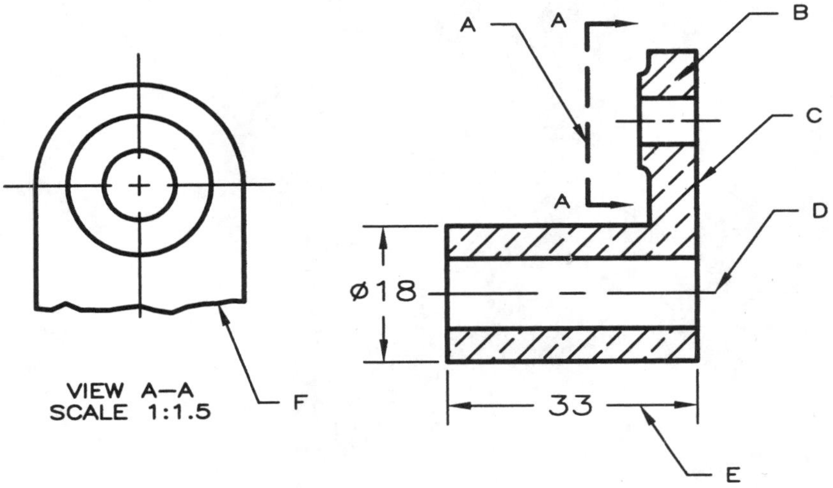

VIEW A—A
SCALE 1:1.5

Ø18

33

1. SKETCH A CIRCLE IN THE SQUARE.
2. SKETCH A CIRCLE USING THE CENTER LINES.
3. SKETCH AN ELLIPSE IN THE RECTANGLE.
4. SKETCH AN ISOMETRIC ELLIPSE IN EACH RHOMBUS.
5. SKETCH THE OBJECTS SHOWN USING THE GRID PROVIDED.

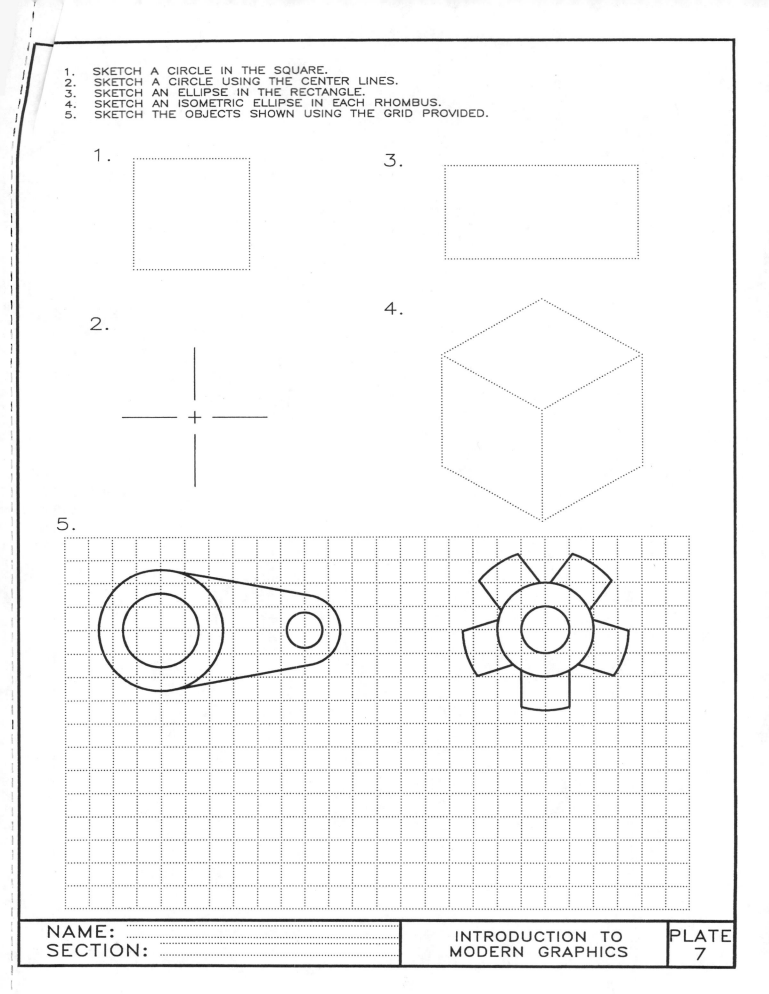

1.

3.

2.

4.

5.

SKETCH EACH OF THE DRAWINGS OF THE BASE PLATE SHOWN BELOW IN THE
GRID SPACE PROVIDED.

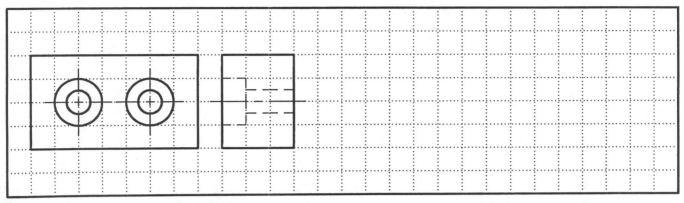

1.   MULTIVIEW PROJECTION

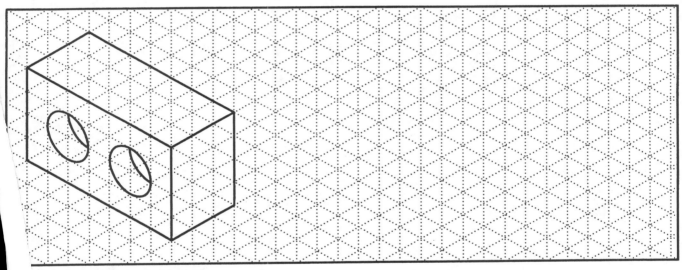

2.   ISOMETRIC PROJECTION

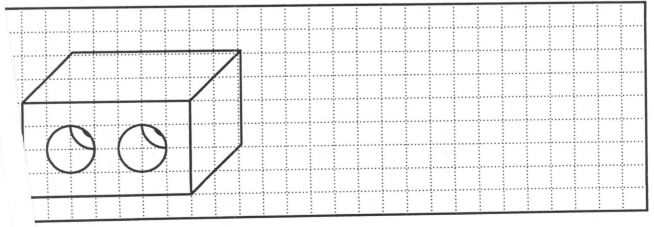

CAVALIER OBLIQUE

SKETCH OR CONSTRUCT THE SYMMETRICAL DRAWING OF THE CHESS PIECE BELOW.

CREATE YOUR OWN SYMMETRICAL SKETCH OF A HOUSEHOLD OBJECT IN THE SPACE BELOW.

ARCHITECTS' SCALE

MEASURE LINES A AND B USING THE FOLLOWING SCALES. LETTER EACH DIMENSION.

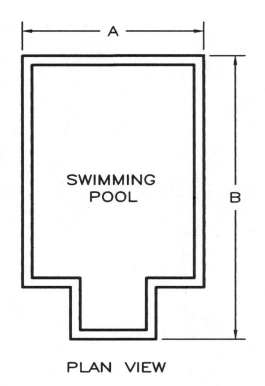

SWIMMING POOL

PLAN VIEW

	A	B
$\frac{1}{4} = 1'-0$	7'-8	
$\frac{3}{32} = 1'-0$		
$1" = 1'-0$		
$\frac{3}{4} = 1'-0$		
Full Size		
$\frac{1}{2}" = 1'-0$		
$1" = 2"$		
$3" = 1'-0$		

ENGINEERS' SCALE

MEASURE C AND D BELOW USING THE APPROPRIATE SCALES. MEASURE ANGLES W, X, Y, AND Z. LETTER EACH DIMENSION.

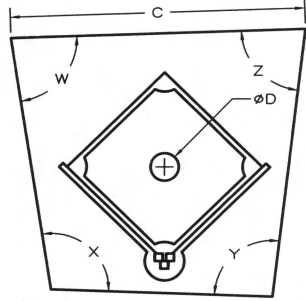

BASEBALL PARK

	C	D
$1" = 100'$	150'	
$1" = 20'$		
$1" = 4'$		
$1" = 30'$		
$1" = 5000'$		

PROTRACTOR

W= ....   Y= ....

X= ....   Z= ....

ARCHITECTS' SCALE
MEASURE LINES A AND B USING THE FOLLOWING SCALES. LETTER EACH DIMENSION.

	A	B
$\frac{1}{8}" = 1'-0$	13'—10	
$\frac{3}{16}" = 1'-0$		
$\frac{1}{4}" = 1'-0$		
$\frac{3}{8}" = 1'-0$		
Full Size		
$1\frac{1}{2}" = 1'-0$		
$1" = 1'-0$		
$3" = 1'-0$		

DOOR

ENGINEERS' SCALE
MEASURE LINES C AND D BELOW USING THE FOLLOWING SCALES. MEASURE ANGLES W, X, Y, AND Z. LETTER EACH DIMENSION.

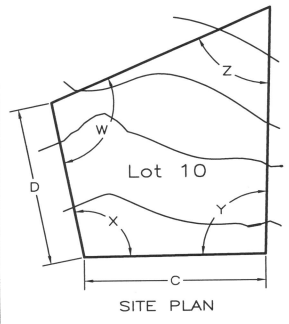

Lot 10

SITE PLAN

	E	F
$1" = 10'$	19'	
$1" = 400'$		
$1" = 5000'$		
$1" = 20'$		
$1" = 30'$		

PROTRACTOR

W=   Y=

X=   Z=

USE THE METRIC SCALE AND MEASURE THE DISTANCES A, B, C, AND R IN THE
TWO—VIEW DRAWING SHOWN BELOW.   BE SURE TO LETTER YOUR ANSWER.

	A(mm)	B(mm)	C(cm)	R(cm)
1. I:I	....	....	....	....
2. I:2	....	....	....	....
3. I:5	....	....	....	....

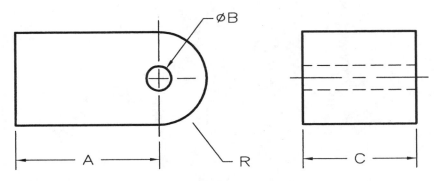

DRAW THE FOLLOWING LINES LISTED BELOW USING THE SPECIFIED SCALES.
A STARTING POINT IS GIVEN FOR EACH PROBLEM.

4. 71mm   SCALE I:I   ⊢

5. 2.3cm   SCALE I:I   ⊢

6. 96mm   SCALE I:2   ⊢

7. 5.4cm   SCALE I:2   ⊢

DIMENSION THE BRACKET BELOW.   SCALE I:I

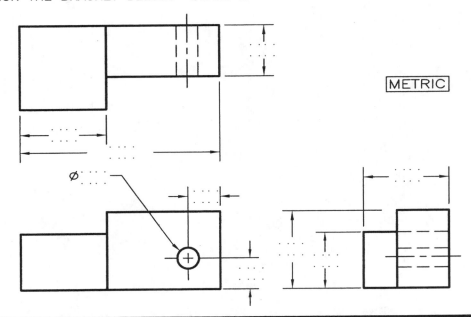

METRIC

1. — MARK THE LOCATION OF A GHOST TOWN THAT IS ONE HALF OF THE DISTANCE
FROM THE CENTER OF STAR AND THE CENTER OF MULESHOE AND MARK WITH A "G".

— THE TOWN YOU JUST LOCATED HAS THREE WORKING GOLD MINES THAT ARE LOCATED
AT EQUAL DISTANCES FROM THE TOWN CENTER. BASED ON THIS INFORMATION,
IDENTIFY THE MINES THAT ARE PLAYED OUT AND LABEL THESE WITH A "P".

— IF THE JOURNEY FROM THE WORKING MINE CLOSEST TO MULESHOE
TO MULESHOE TAKES FIVE EQUAL TRAVEL DAYS, MARK THE DISTANCE TRAVELED
AT THE END OF THE FOURTH DAY WITH A SHORT LINE LABELED "T".

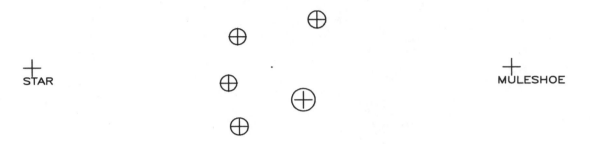

2. — THE FOLLOWING DRAWING IS A SITE LAYOUT FOR A ROAD.
COMPLETE THE LARGE CURVED CURVED CORNER. THE INSIDE
RADIUS OF THE CURVE IS 20 FEET. MARK ALL TANGENT POINTS.

— THE OWNERS WOULD LIKE AN OAK TREE PLACED AT THE END
OF A DIAGONAL LINE ALONG THE BISECTOR OF THE ANGLE FORMED BY
THE HEDGE. THIS TREE IS 55 FEET FROM THE INTERSECTION OF THE HEDGES.
MARK THE LOCATION OF THIS TREE WITH A "X".

— COMPLETE THE REMAINING PARTS OF THE ROAD USING AN INSIDE RADIUS OF 5 FEET.
MARK ALL TANGENT POINTS.

SCALE 1"= 20'

| NAME: .... | INTRODUCTION TO | PLATE |
| SECTION: .... | MODERN GRAPHICS | 17 |

SKETCH THE MULTIVIEWS FOR THE PROBLEMS SHOWN BELOW.

1.

2.

3.

4.

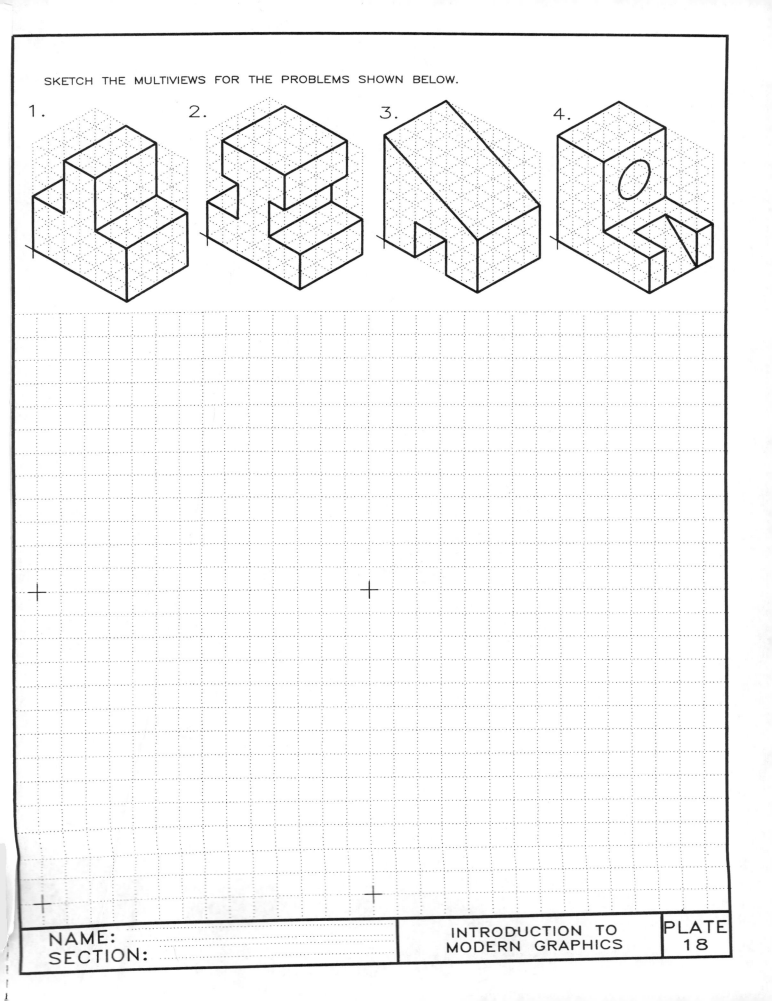

SKETCH THE MULTIVIEWS FOR THE PROBLEMS SHOWN BELOW.

1.

2.

3.

4.

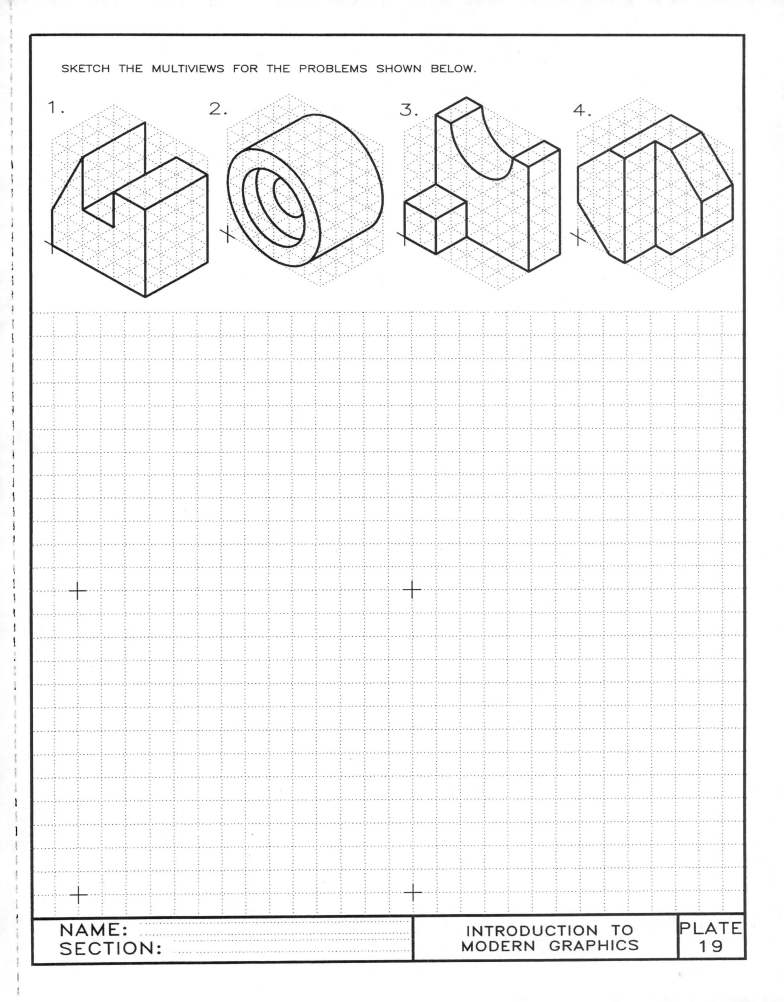

NAME: ....

SECTION: ....

INTRODUCTION TO
MODERN GRAPHICS

PLATE
19

SKETCH THE NECESSARY MULTIVIEWS FOR THE PROBLEMS SHOWN BELOW.

1.

2.

3.

4.

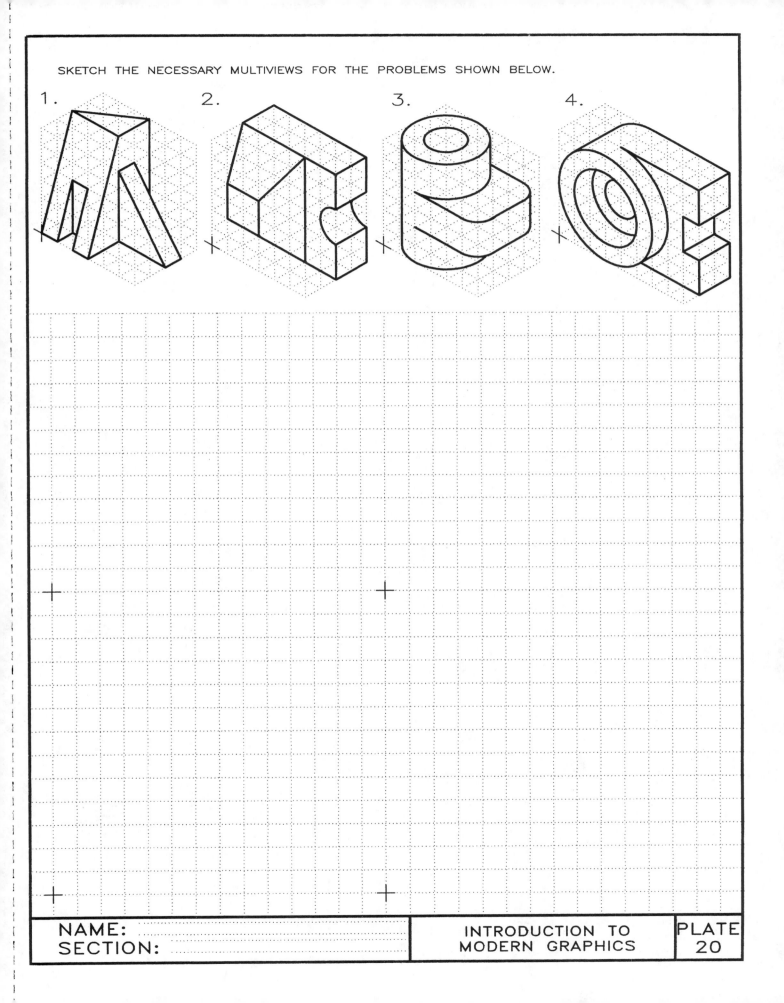

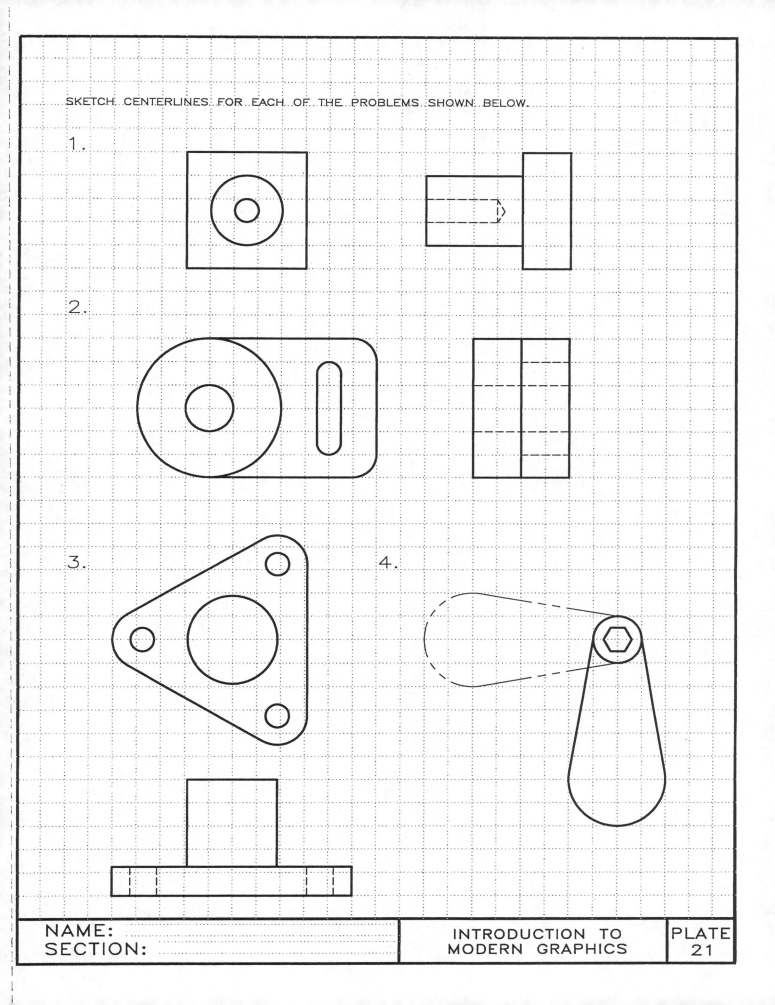

SKETCH CENTERLINES FOR EACH OF THE PROBLEMS SHOWN BELOW.

1.

2.

3.   4.

NAME:
SECTION:

INTRODUCTION TO
MODERN GRAPHICS

PLATE
21

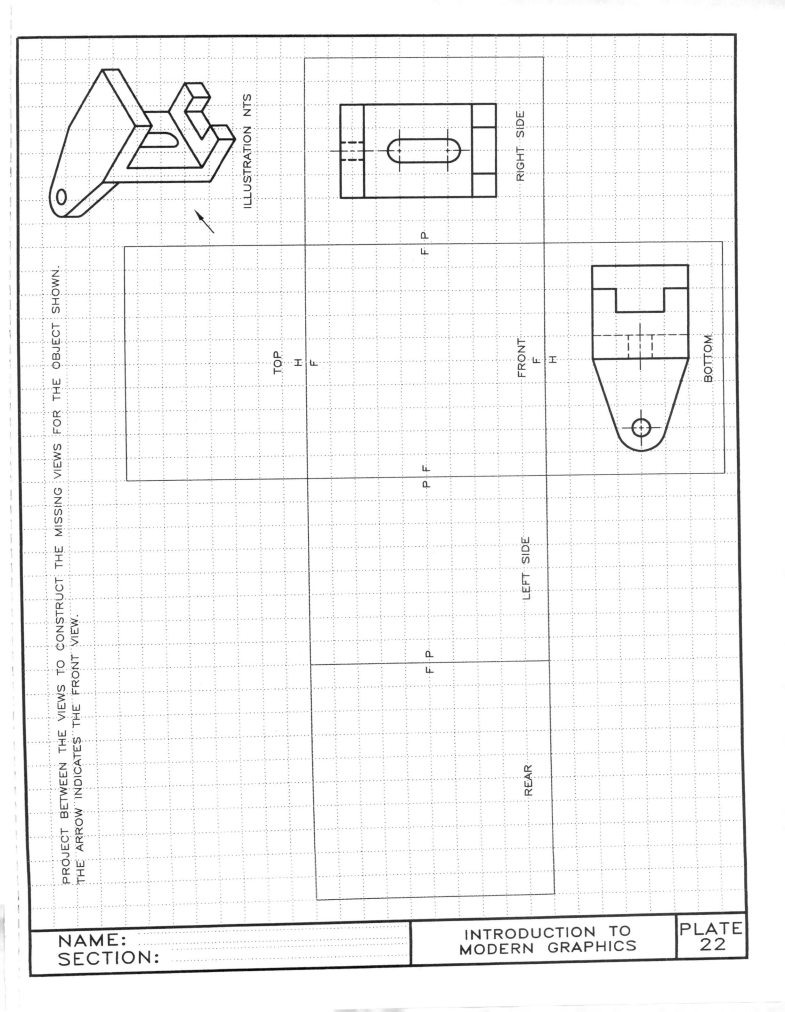

ILLUSTRATION NTS

PROJECT BETWEEN THE VIEWS TO CONSTRUCT THE MISSING VIEWS FOR THE OBJECT SHOWN.
THE ARROW INDICATES THE FRONT VIEW.

RIGHT SIDE

TOP

FRONT

BOTTOM

LEFT SIDE

REAR

F P

H F

F H

P F

F P

F P

NAME:
SECTION:

INTRODUCTION TO
MODERN GRAPHICS

PLATE
22

FOR PROBLEMS 1 AND 2 CONSTRUCT FOUR DIFFERENT FRONT VIEWS THAT WORK AS
POSSIBLE SOLUTIONS FOR THE TOP VIEW GIVEN. FOR PROBLEM 3 CONSTRUCT A FRONT VIEW
THAT WORK AS A SOLUTION WITH THE TOP AND RIGHT SIDE VIEW GIVEN FOR EACH OF THE
PROBLEMS.

1.

2.

3.

COMPLETE THE CHART BELOW BY MARKING AN X IN THE
APPROPRIATE COLUMNS.

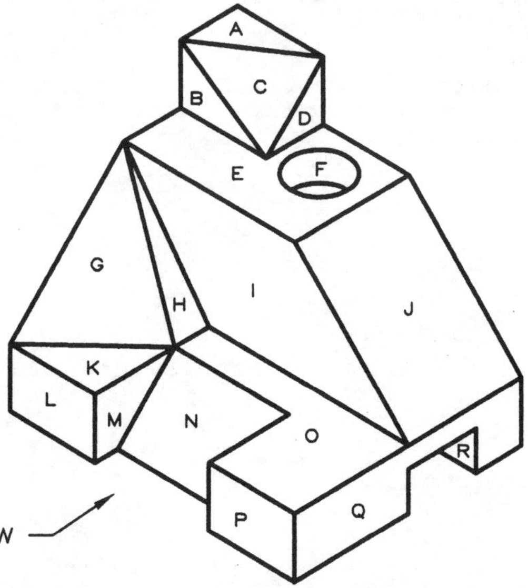

FRONT VIEW ⟶

	A	B	C	D	E	F	G	H	I	J	K	L	M	N	O	P	Q	R
1. NORMAL SURFACES																		
2. INCLINED SURFACES																		
3. OBLIQUE SURFACES																		
4. SURFACES THAT APPEAR HIDDEN IN THE FRONT VIEW																		
5. SURFACES THAT ARE VISIBLE EDGES IN THE RIGHT SIDE VIEW																		
6. SURFACES THAT ARE HIDDEN EDGES IN THE RIGHT SIDE VIEW																		
7. SURFACES THAT ARE VISIBLE EDGES IN THE FRONT VIEW																		
8. SURFACES THAT ARE VISIBLE EDGES IN THE TOP VIEW																		
9. SURFACES FORESHORTENED WHEN DRAWN AS MULTIVIEWS																		

	FRONT	TOP	RIGHT	REAR	BOTTOM	LEFT
10. VIEWS WHERE THE HOLE APPEARS AS HIDDEN LINES						
11. VIEWS WHERE THE HOLE APPEARS AS A VISIBLE CIRCLE						
12. VIEWS WHERE THE HOLE APPEARS AS A HIDDEN CIRCLE						
13. VIEWS NEEDING HIDDEN LINES						
14. VIEWS NEEDING CENTER MARKS						
15. VIEWS NEEDING CENTER LINES						

NAME:
SECTION:

INTRODUCTION TO
MODERN GRAPHICS

PLATE
25

EACH OF THE FOLLOWING DRAWINGS IS INCOMPLETE. PLEASE ADD THE NEEDED MISSING LINES TO COMPLETE THE VIEWS. YOU MAY NOT ADD ANY LINES OUTSIDE OF THE EDGES OF THE PRESENT VIEWS. THE LINES MAY BE HIDDEN LINES OR OBJECT LINES. PLEASE ADD CENTERLINES WHERE NEEDED. IT IS OFTEN HELPFUL TO SKETCH THE OBJECT AND/OR TO NUMBER THE POINTS.

1

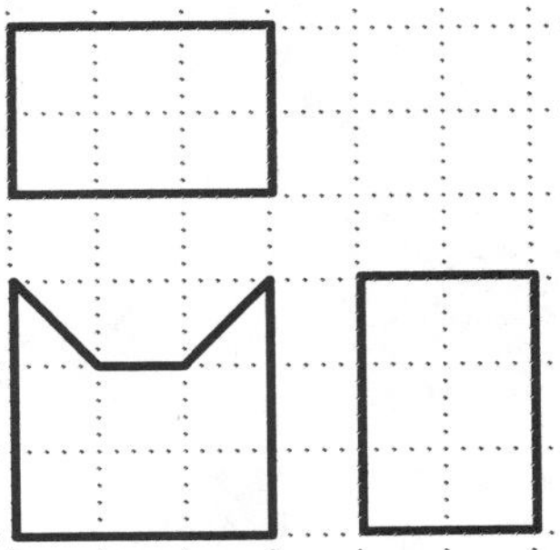

2

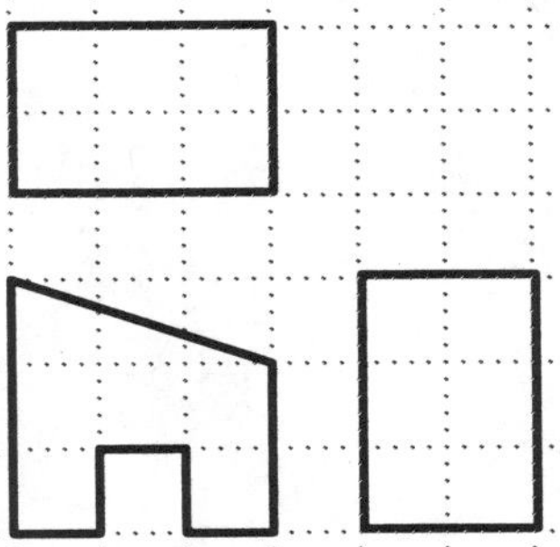

3

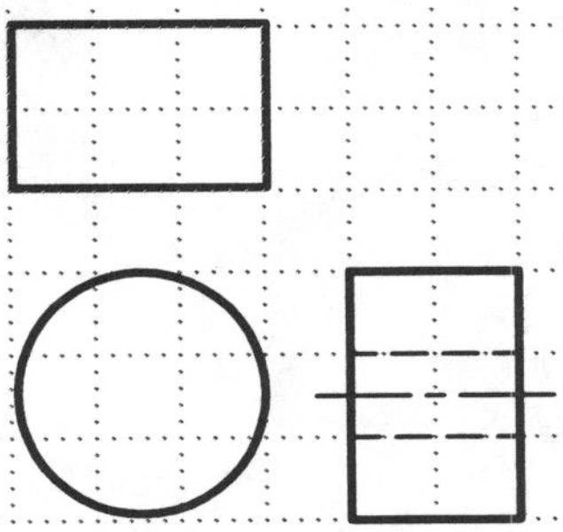

4

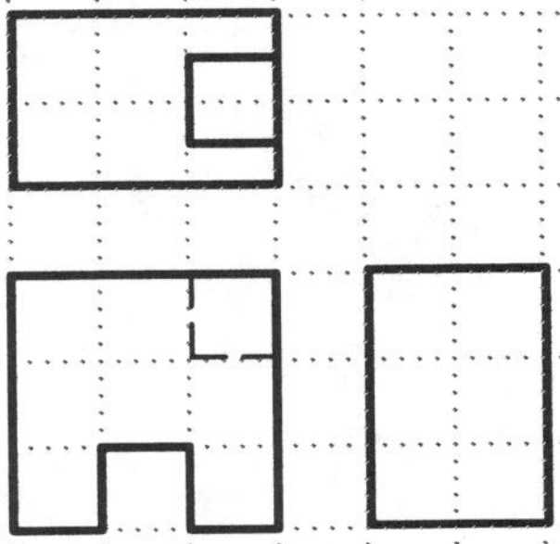

5

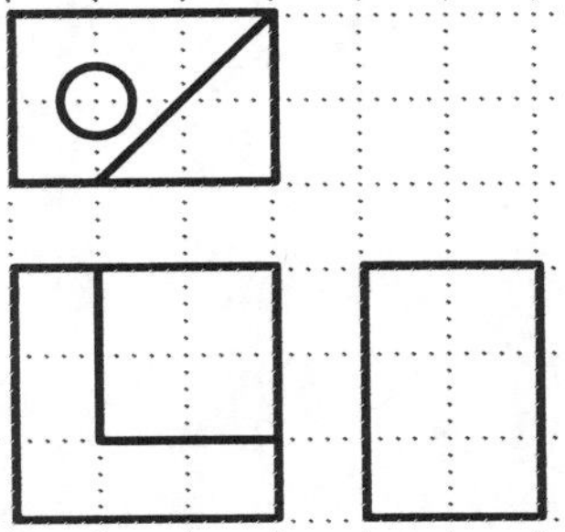

6

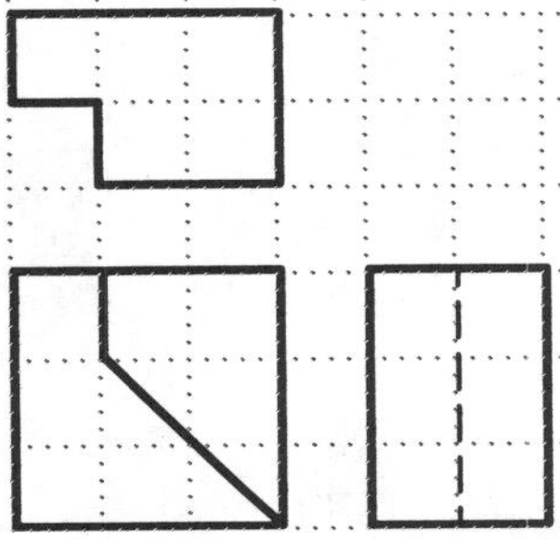

7

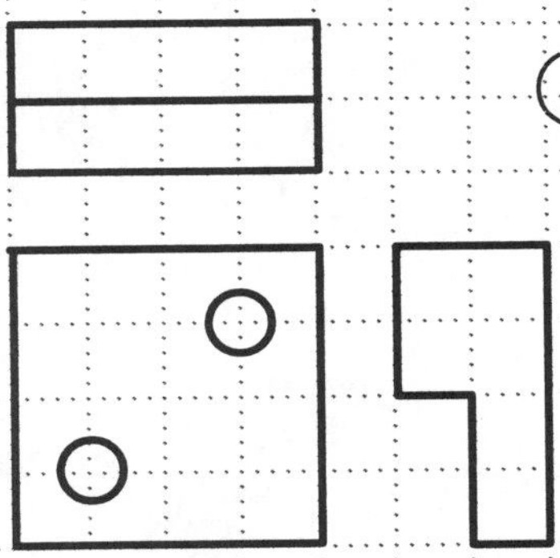

8

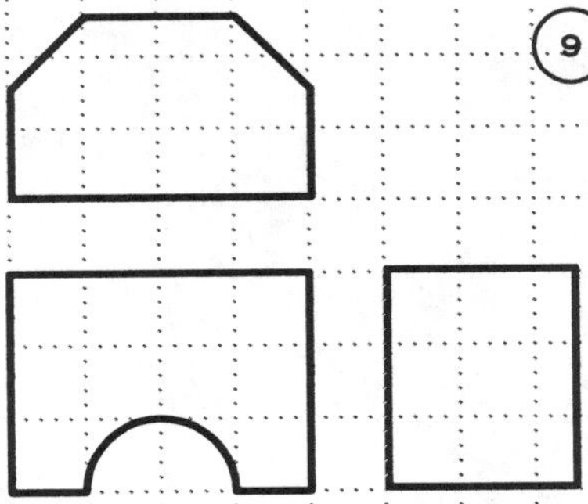

9

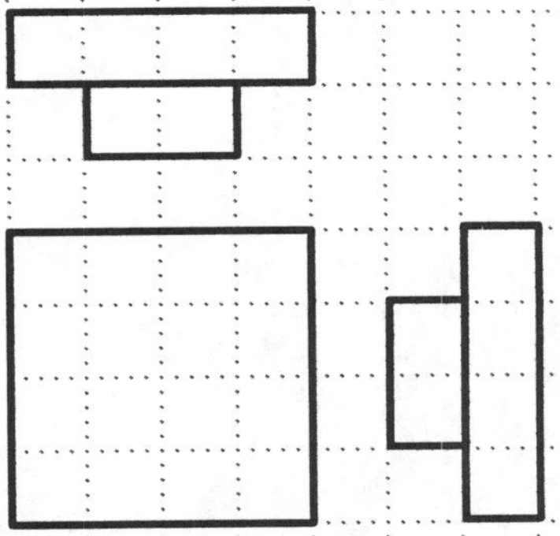

10

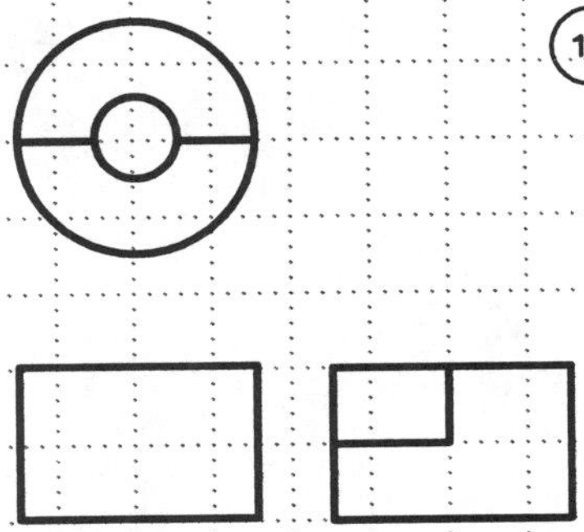

11

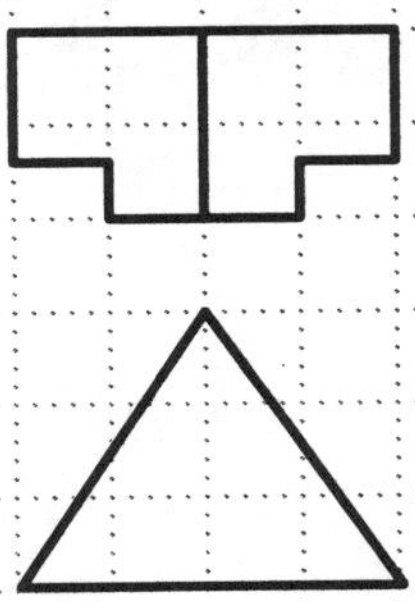

12

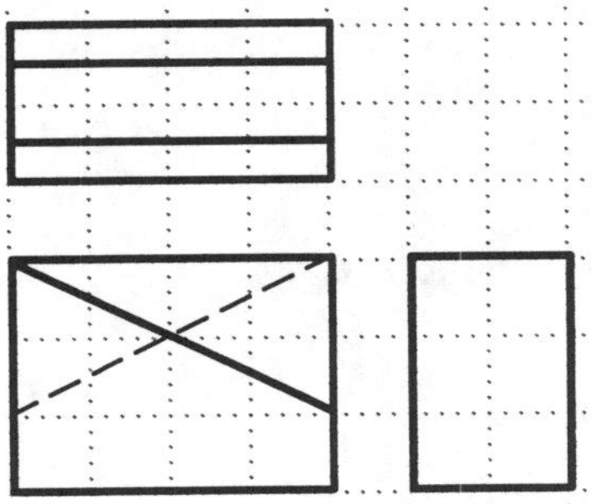

NAME: SECTION:	INTRODUCTION TO MODERN GRAPHICS	PLATE 26

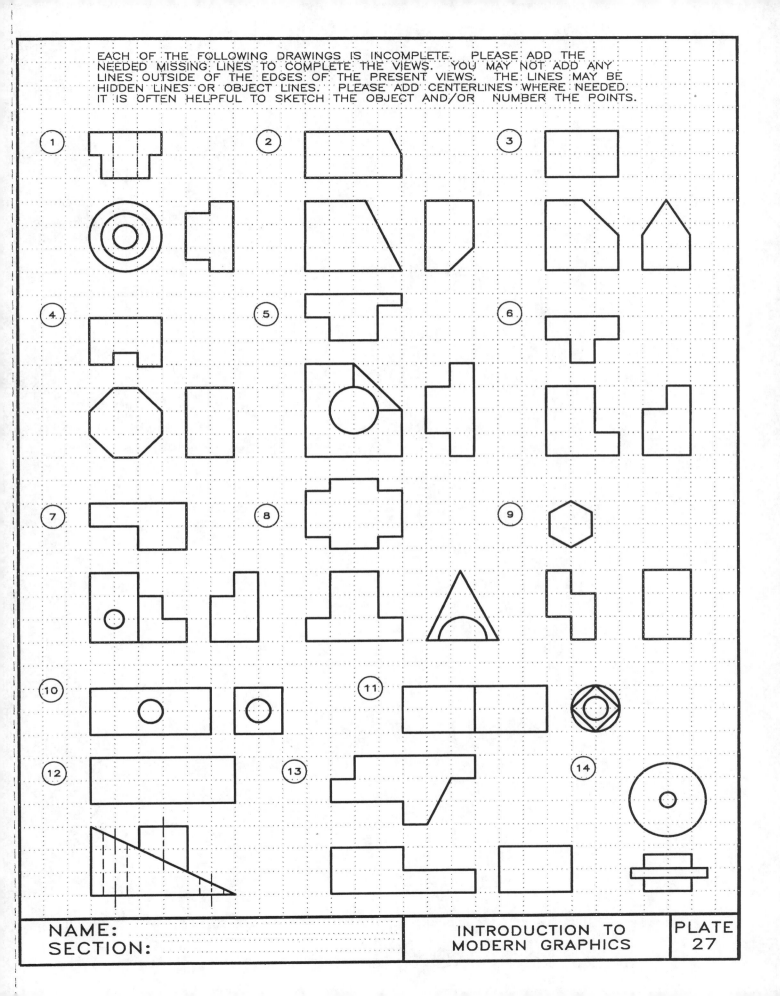

EACH OF THE FOLLOWING DRAWINGS IS INCOMPLETE. PLEASE ADD THE
NEEDED MISSING LINES TO COMPLETE THE VIEWS. YOU MAY NOT ADD ANY
LINES OUTSIDE OF THE EDGES OF THE PRESENT VIEWS. THE LINES MAY BE
HIDDEN LINES OR OBJECT LINES. PLEASE ADD CENTERLINES WHERE NEEDED.
IT IS OFTEN HELPFUL TO SKETCH THE OBJECT AND/OR NUMBER THE POINTS.

NAME:
SECTION:

INTRODUCTION TO
MODERN GRAPHICS

PLATE
27

COMPLETE THE THREE VIEWS AND THE ISOMETRIC FOR EACH OF THE GIVEN OBJECTS.

1.

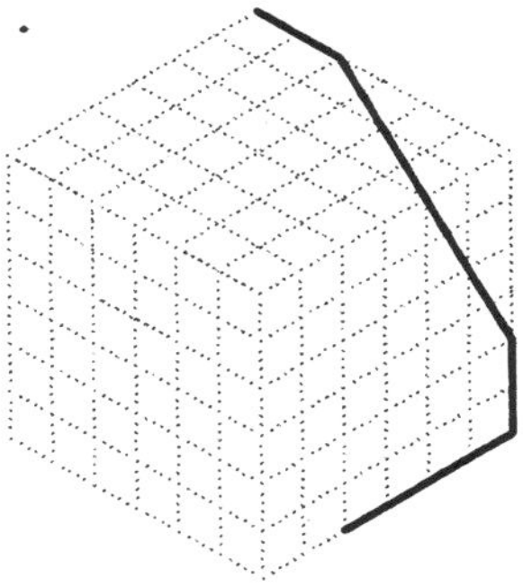

2.

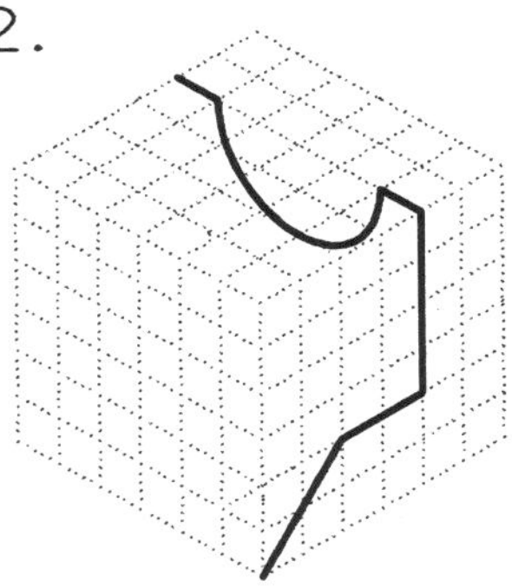

3.

4.

(1) WEDGE BLOCK

(2) SHAFT SUPPORT

(3) SPACER

(4) SADDLE

NAME: SECTION:	INTRODUCTION TO MODERN GRAPHICS	PLATE 28

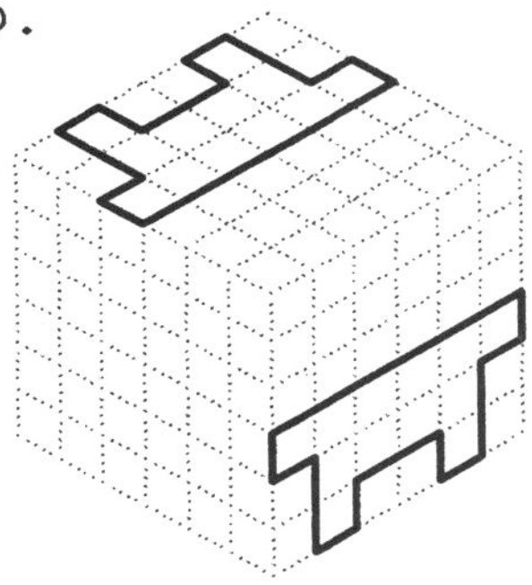

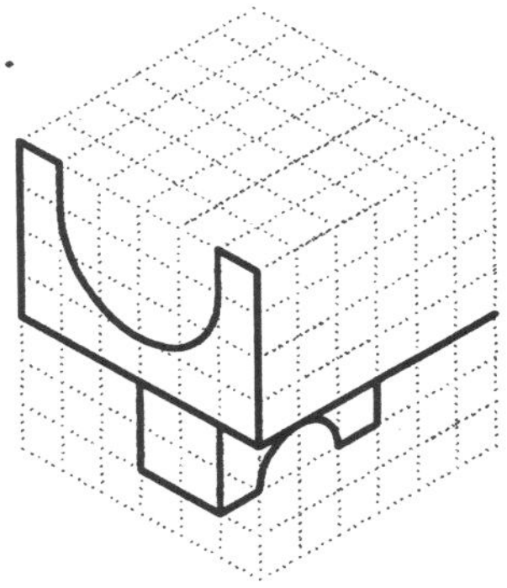

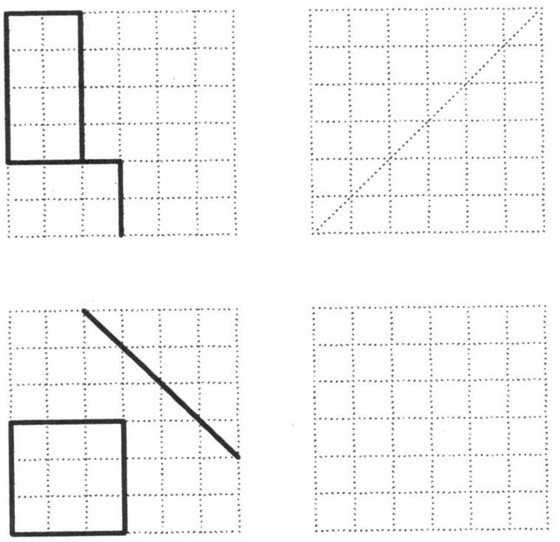

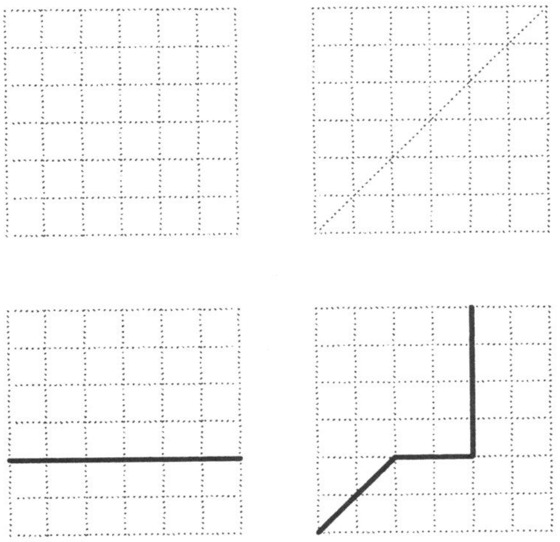

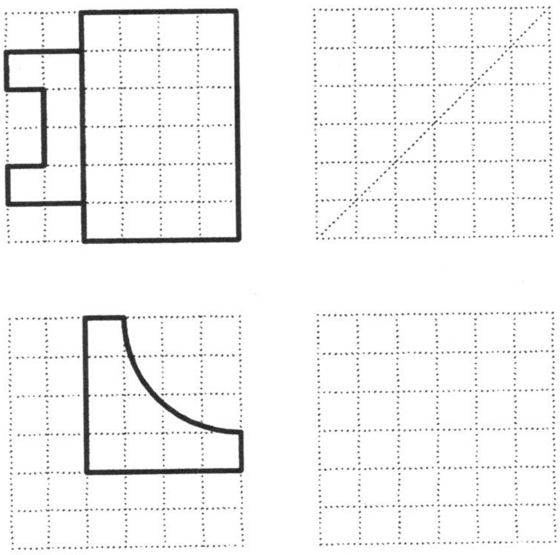

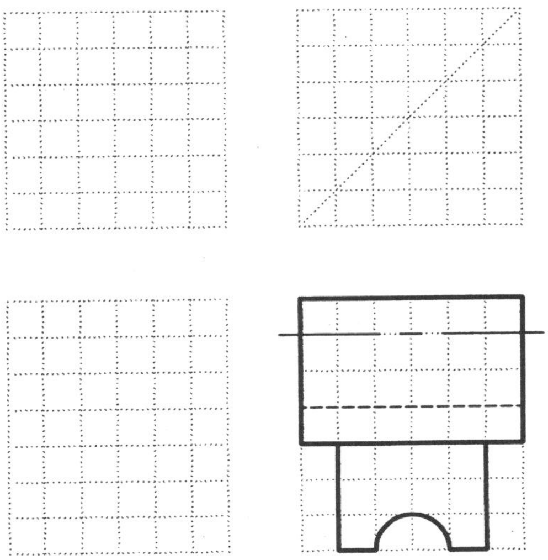

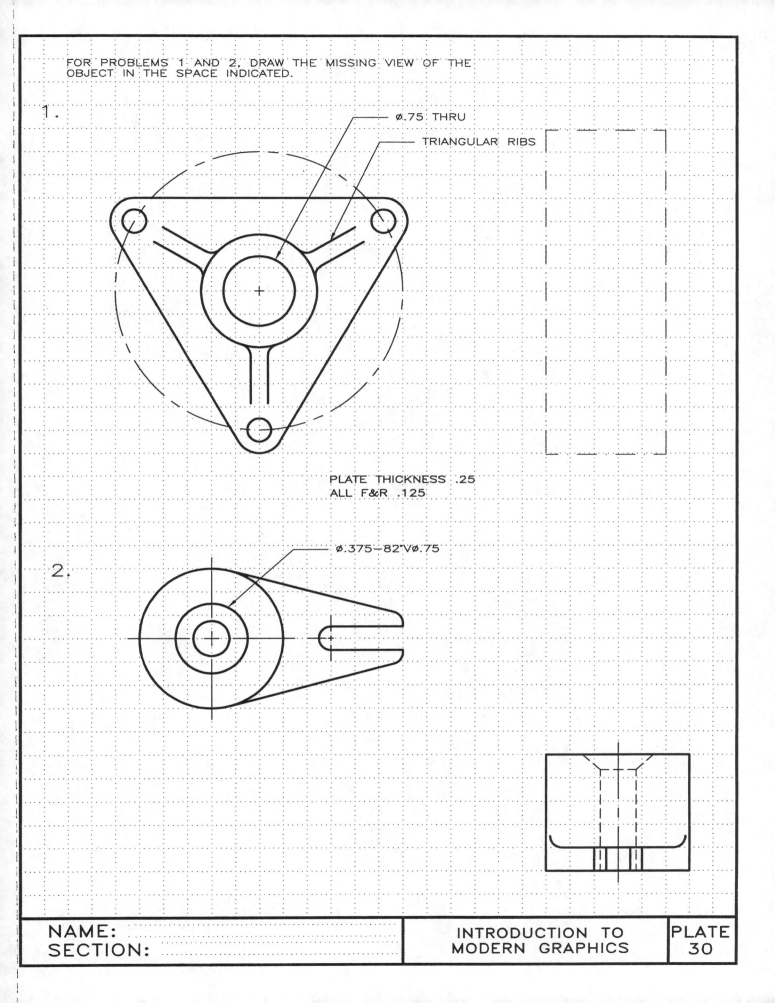

FOR PROBLEMS 1 AND 2, DRAW THE MISSING VIEW OF THE
OBJECT IN THE SPACE INDICATED.

1.

Ø.75 THRU

TRIANGULAR RIBS

PLATE THICKNESS .25
ALL F&R .125

Ø.375−82°VØ.75

2.

SKETCH 3 VIEWS OF THE BEARING.

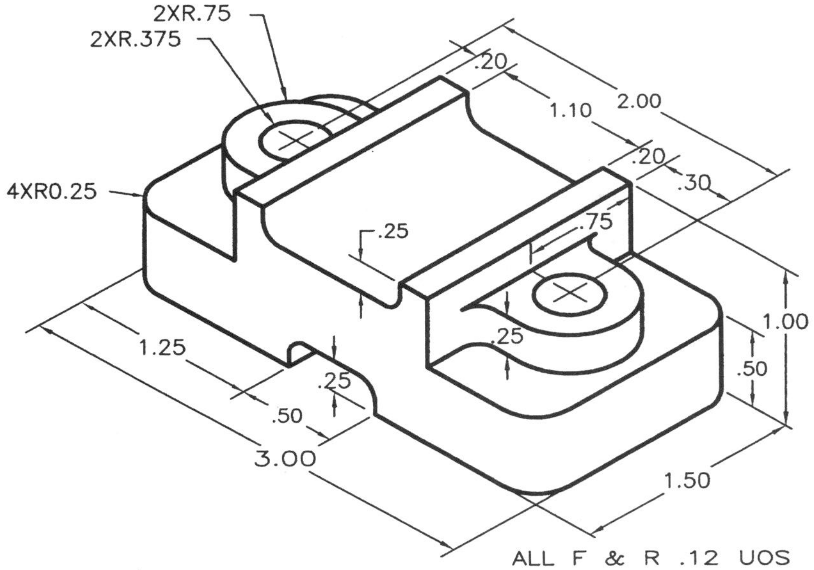

2XR.75
2XR.375
4XR0.25
.20
1.10
2.00
.20
.30
.25
.75
1.00
.25
.50
1.25
.25
.50
3.00
1.50
ALL F & R .12 UOS

NAME: SECTION:	INTRODUCTION TO MODERN GRAPHICS	PLATE 31

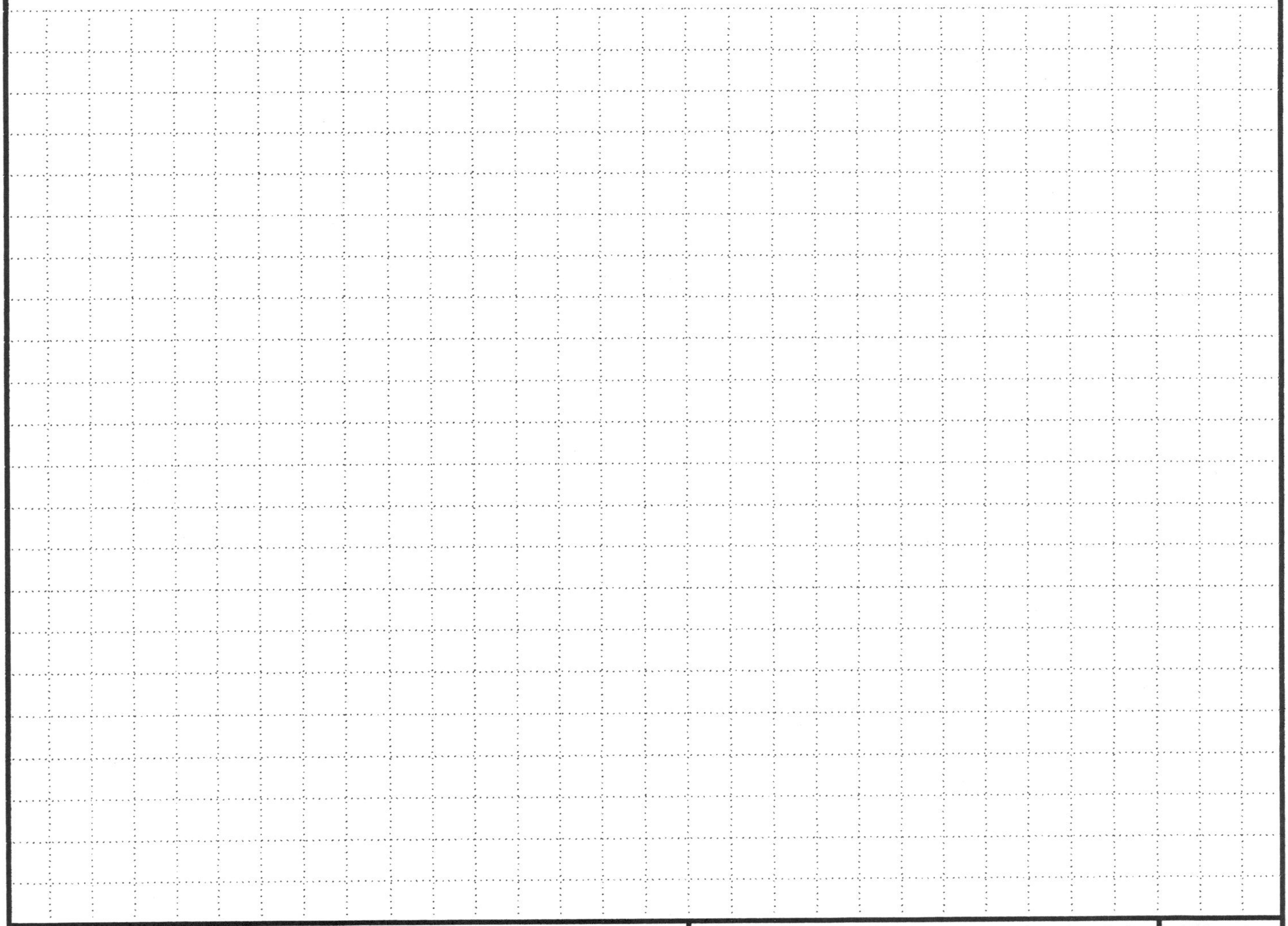

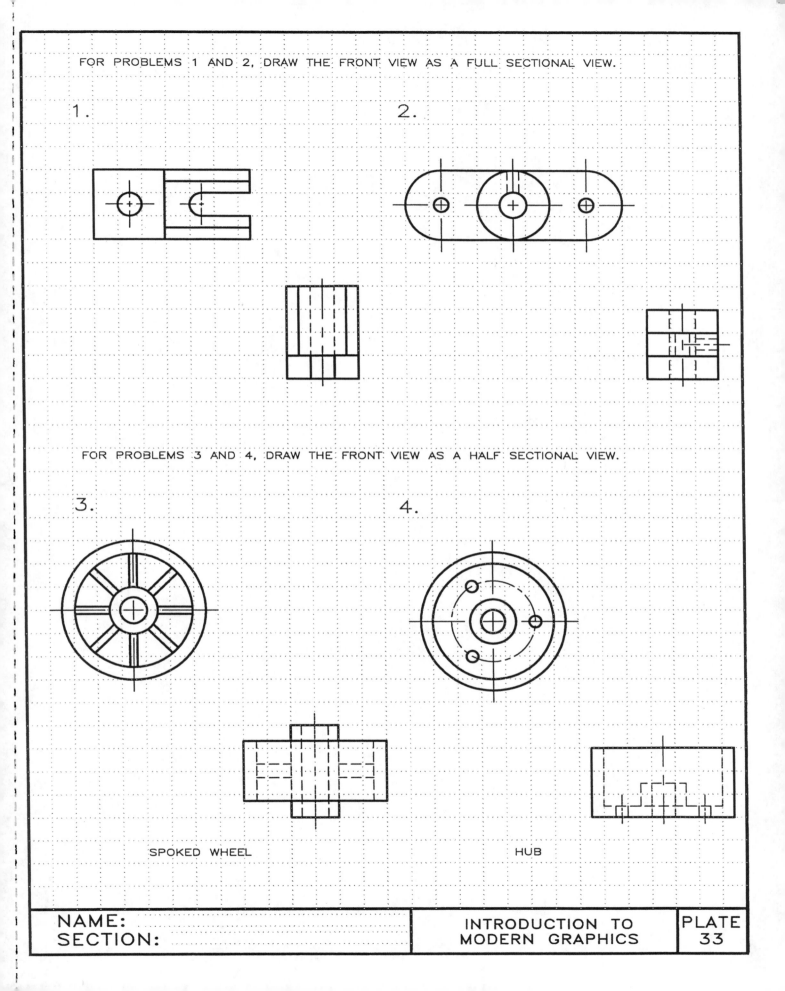

FOR PROBLEMS 1 AND 2, DRAW THE FRONT VIEW AS A FULL SECTIONAL VIEW.

1.

2.

FOR PROBLEMS 3 AND 4, DRAW THE FRONT VIEW AS A HALF SECTIONAL VIEW.

3.

4.

SPOKED WHEEL

HUB

NAME:

SECTION:

INTRODUCTION TO
MODERN GRAPHICS

PLATE
33

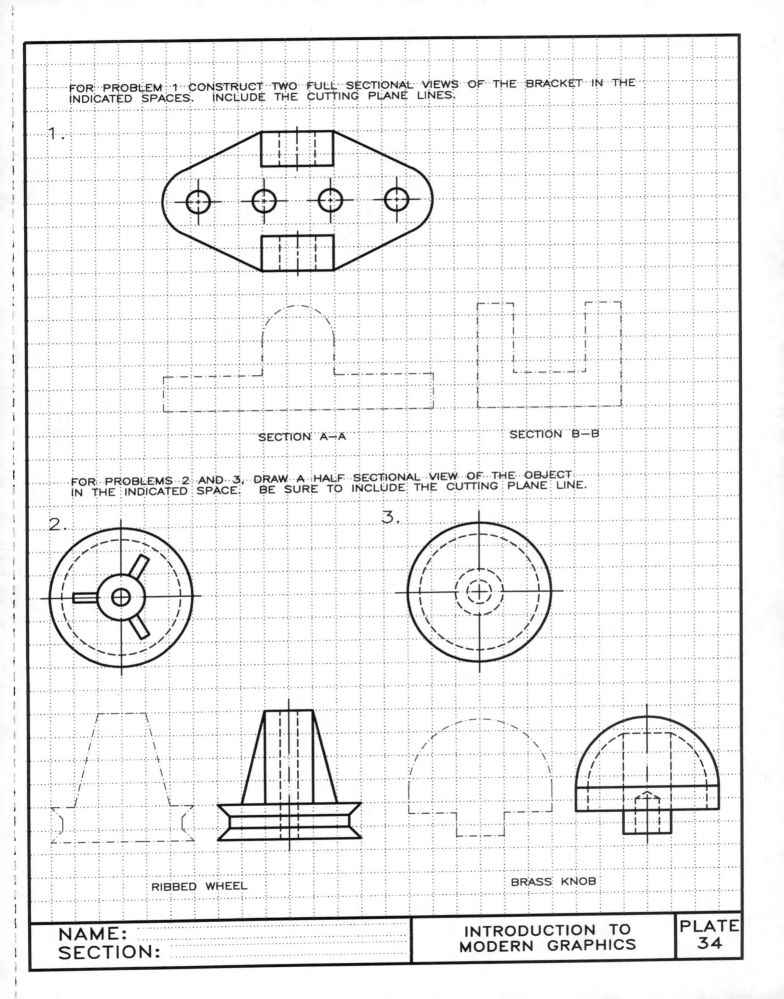

FOR PROBLEM 1 CONSTRUCT TWO FULL SECTIONAL VIEWS OF THE BRACKET IN THE INDICATED SPACES. INCLUDE THE CUTTING PLANE LINES.

1.

SECTION A—A   SECTION B—B

FOR PROBLEMS 2 AND 3, DRAW A HALF SECTIONAL VIEW OF THE OBJECT IN THE INDICATED SPACE. BE SURE TO INCLUDE THE CUTTING PLANE LINE.

2.   3.

RIBBED WHEEL   BRASS KNOB

NAME:
SECTION:

INTRODUCTION TO
MODERN GRAPHICS

PLATE
34

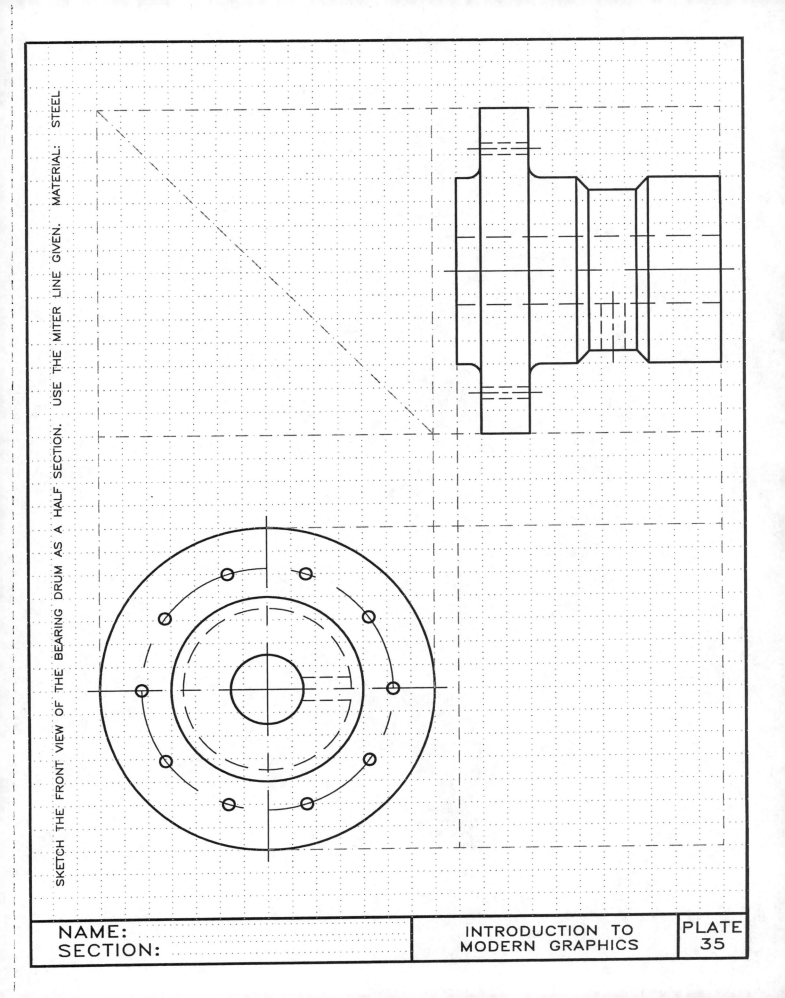

SKETCH THE FRONT VIEW OF THE BEARING DRUM AS A HALF SECTION. USE THE MITER LINE GIVEN. MATERIAL: STEEL

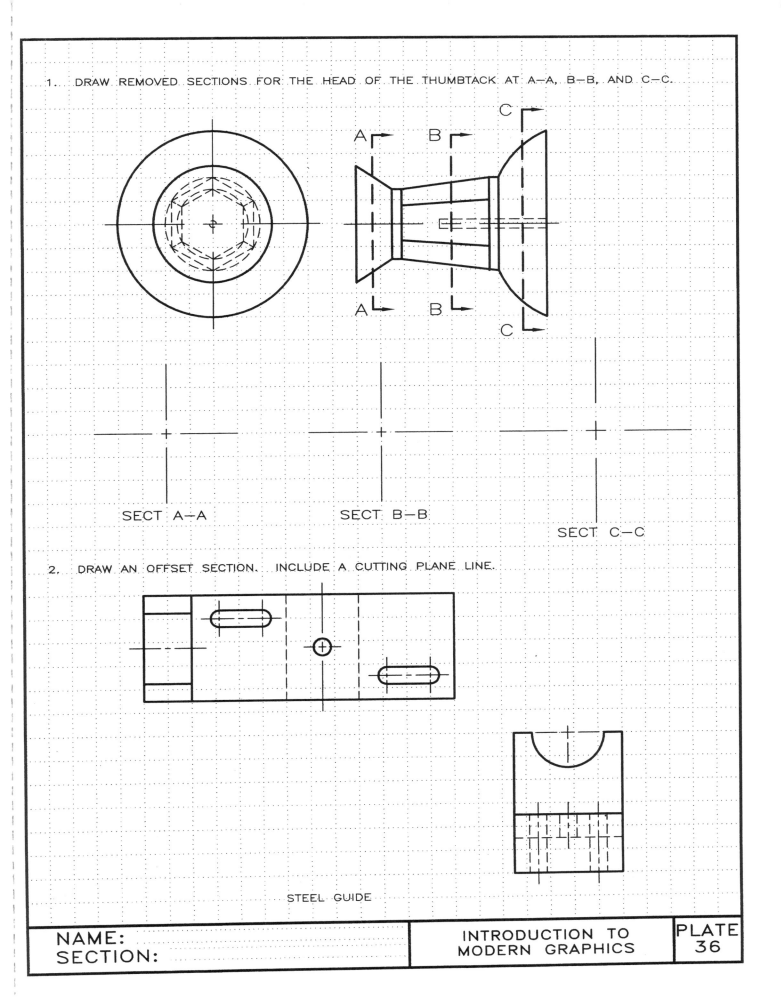

1. DRAW REMOVED SECTIONS FOR THE HEAD OF THE THUMBTACK AT A—A, B—B, AND C—C.

SECT A—A

SECT B—B

SECT C—C

2. DRAW AN OFFSET SECTION. INCLUDE A CUTTING PLANE LINE.

STEEL GUIDE

NAME:

SECTION:

INTRODUCTION TO
MODERN GRAPHICS

PLATE
36

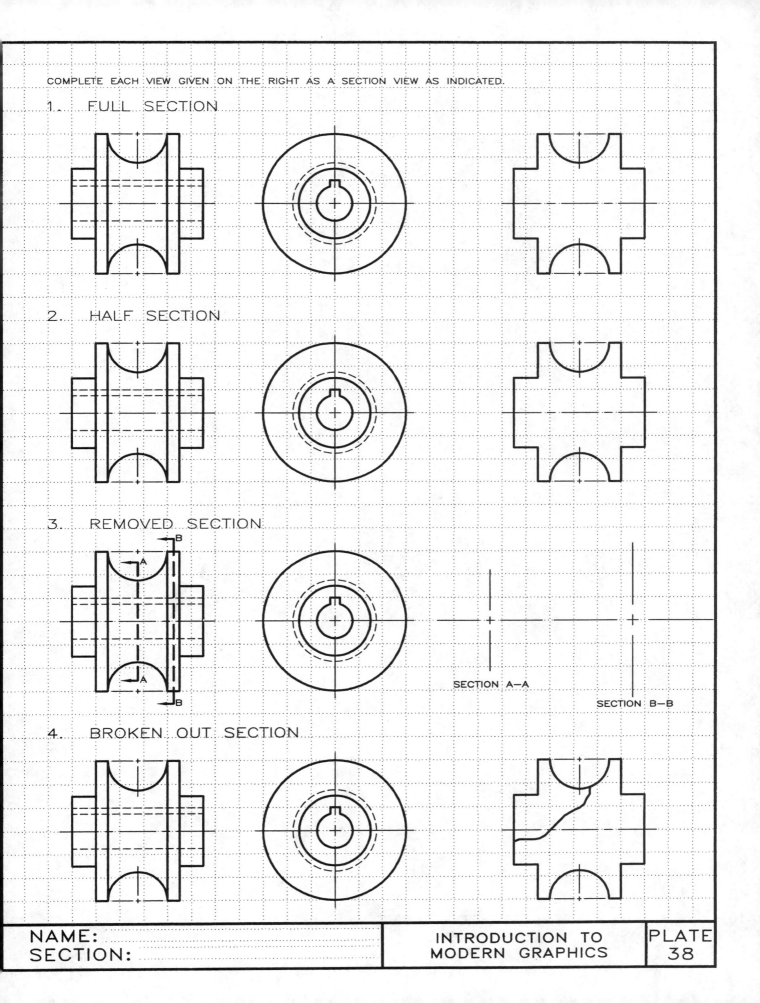

COMPLETE EACH VIEW GIVEN ON THE RIGHT AS A SECTION VIEW AS INDICATED.

1. FULL SECTION

2. HALF SECTION

3. REMOVED SECTION

SECTION A—A

SECTION B—B

4. BROKEN OUT SECTION

NAME:
SECTION:

INTRODUCTION TO
MODERN GRAPHICS

PLATE
38

1.   DETERMINE THE TRUE LENGTH OF THE GIVEN LINES IN THE FRONT VIEW BY REVOLUTION.

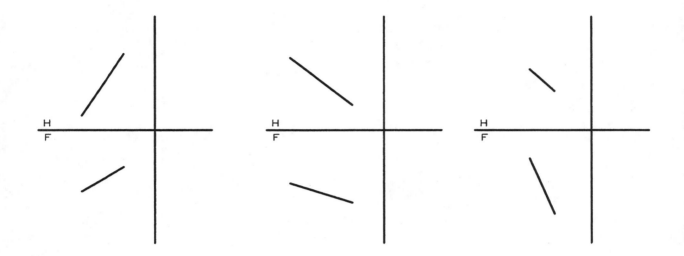

2.   DETERMINE THE TRUE LENGTH OF THE GIVEN LINES BY AUXILIARY VIEW.

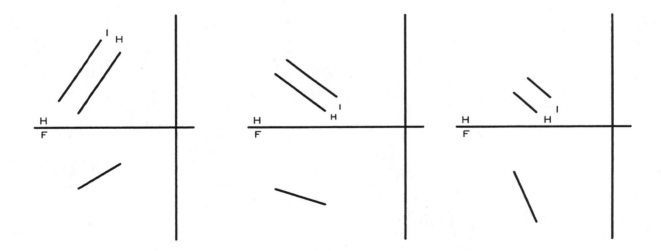

CREATE AN AUXILIARY VIEW OF THE
INDICATED INCLINED SURFACES.

1.

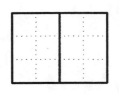

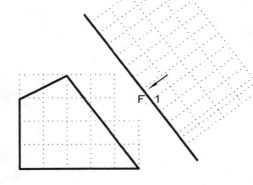

F 1

WHEEL STOP
WOOD

2.

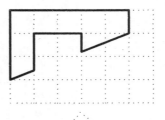

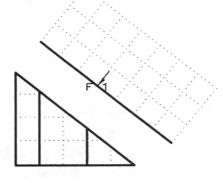

F 1

SLOTTED WEDGE
WOOD

3.

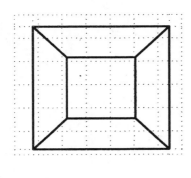

R P

PLANT STAND
WOOD

4.

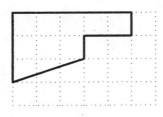

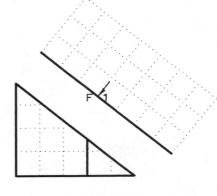

F 1

PUSH STICK
WOOD

NAME: ....
SECTION: ....

INTRODUCTION TO
MODERN GRAPHICS

PLATE
40

DRAW A PARTIAL AUXILIARY VIEW OF THE INDICATED INCLINED SURFACES FOR EACH OBJECT.

1.

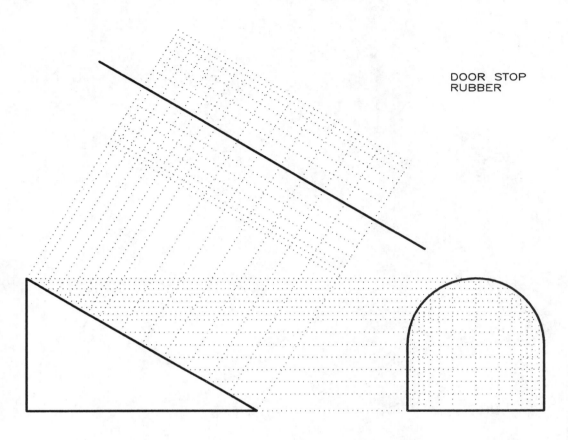

DOOR STOP
RUBBER

2.

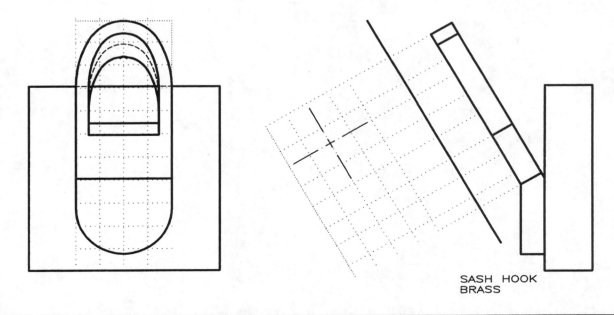

SASH HOOK
BRASS

CREATE AN AN AUXILIARY VIEW OF THE CASCADE FOUNTAIN.

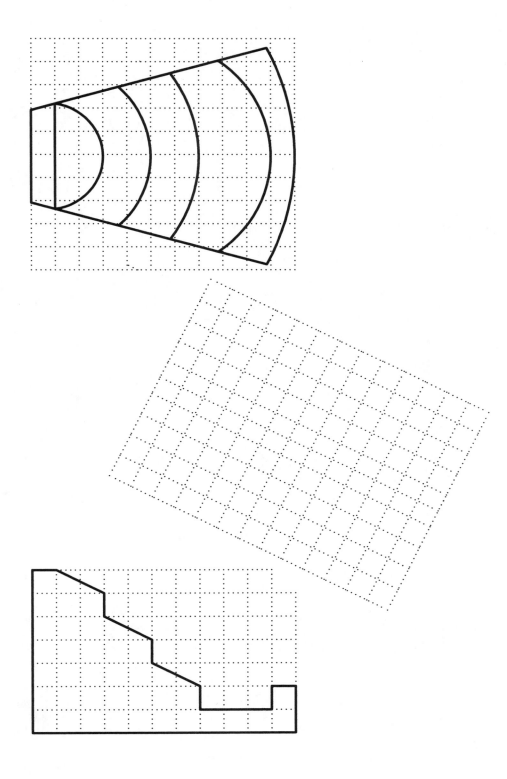

INTRODUCTION TO
MODERN GRAPHICS

PLATE
42

DRAW A DEVELOPMENT LAYOUT FOR EACH OBJECT

1.

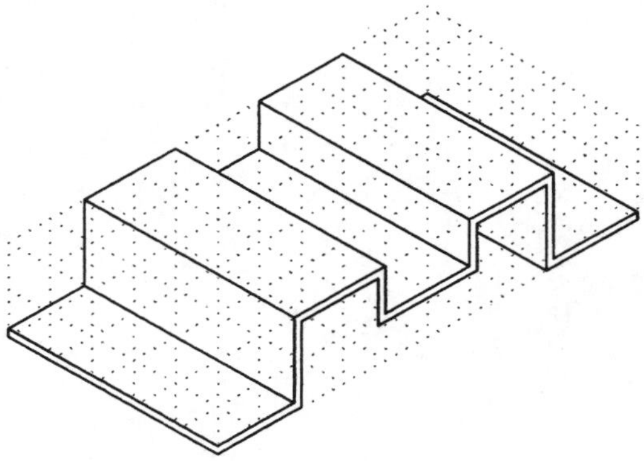

CABLE MOUNT
BRASS

2.

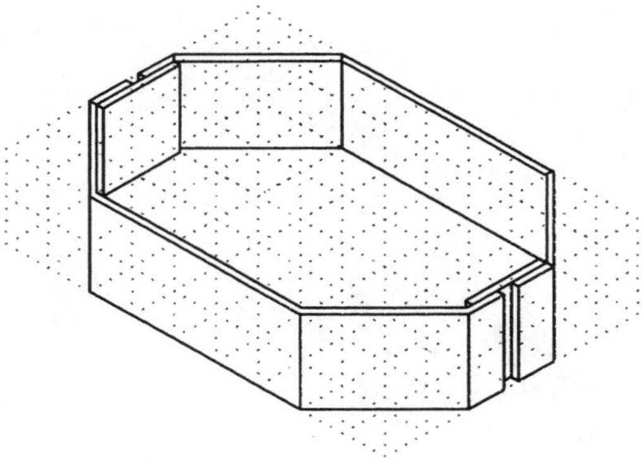

BOX TRAY
COPPER

DRAW A SHEETMETAL DEVELOPMENT FOR THE TOOL BOX AND HANDLE. ALLOW 1/16 FOR THE FOLDED EDGES.

2.00
0.40
0.25
0.80
0.25
2.40
0.83
1.66

NAME:
SECTION:

INTRODUCTION TO MODERN GRAPHICS

PLATE 44

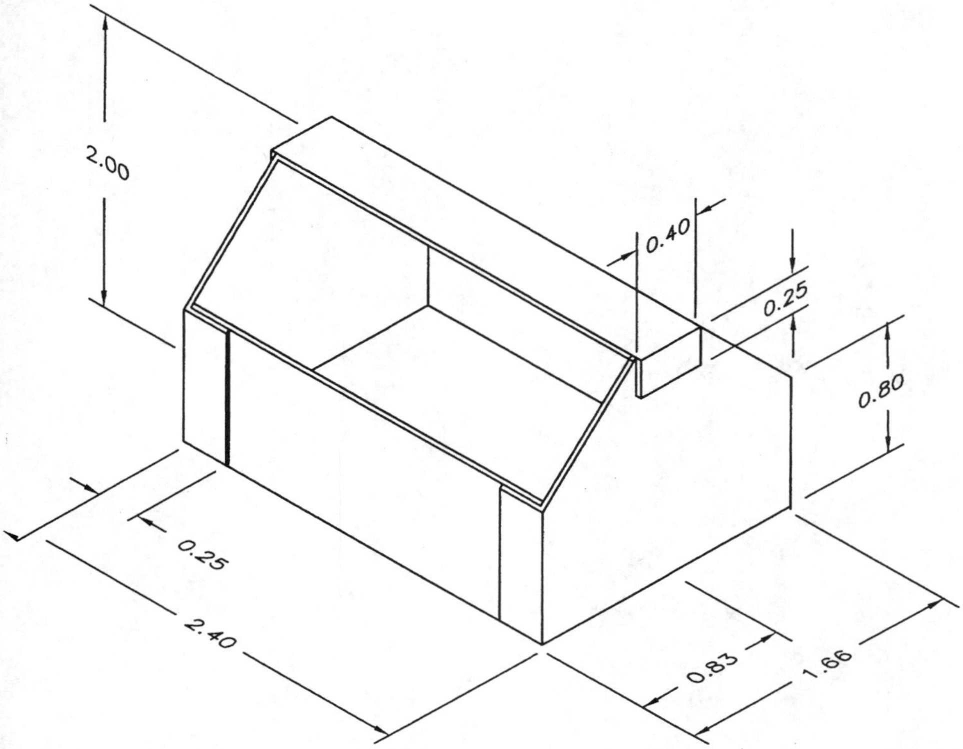

DIMENSION THE FOLLOWING OBJECTS.

1.

2.

DIMENSION THE FOLLOWING OBJECTS.

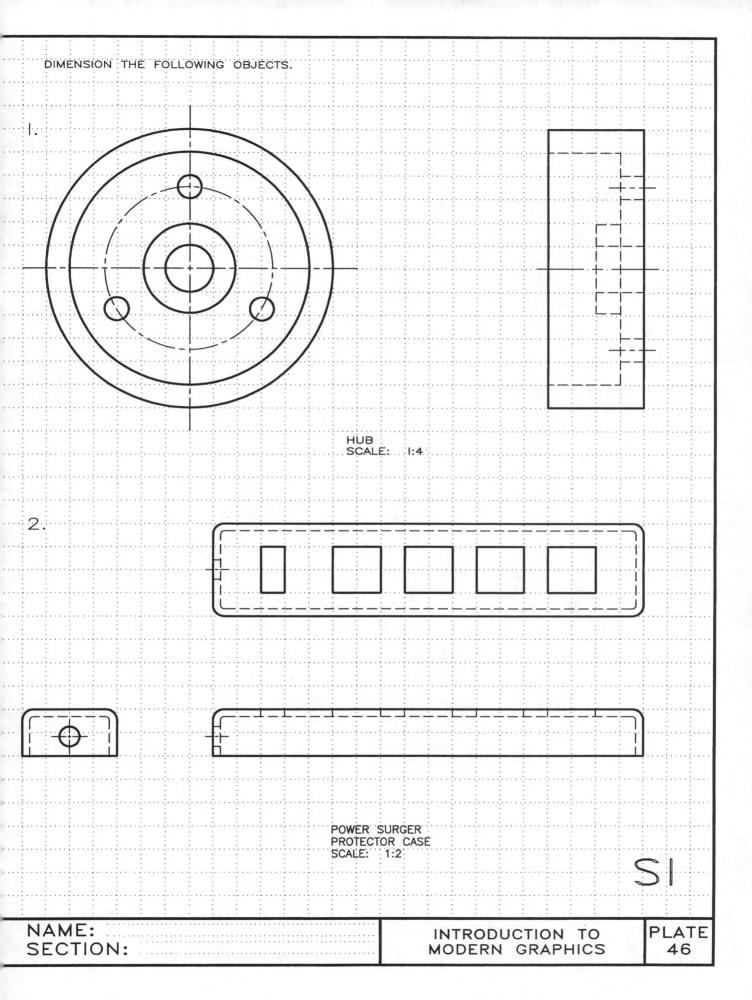

1.

HUB
SCALE: 1:4

2.

POWER SURGER
PROTECTOR CASE
SCALE: 1:2

S1

NAME:
SECTION:

INTRODUCTION TO
MODERN GRAPHICS

PLATE
46

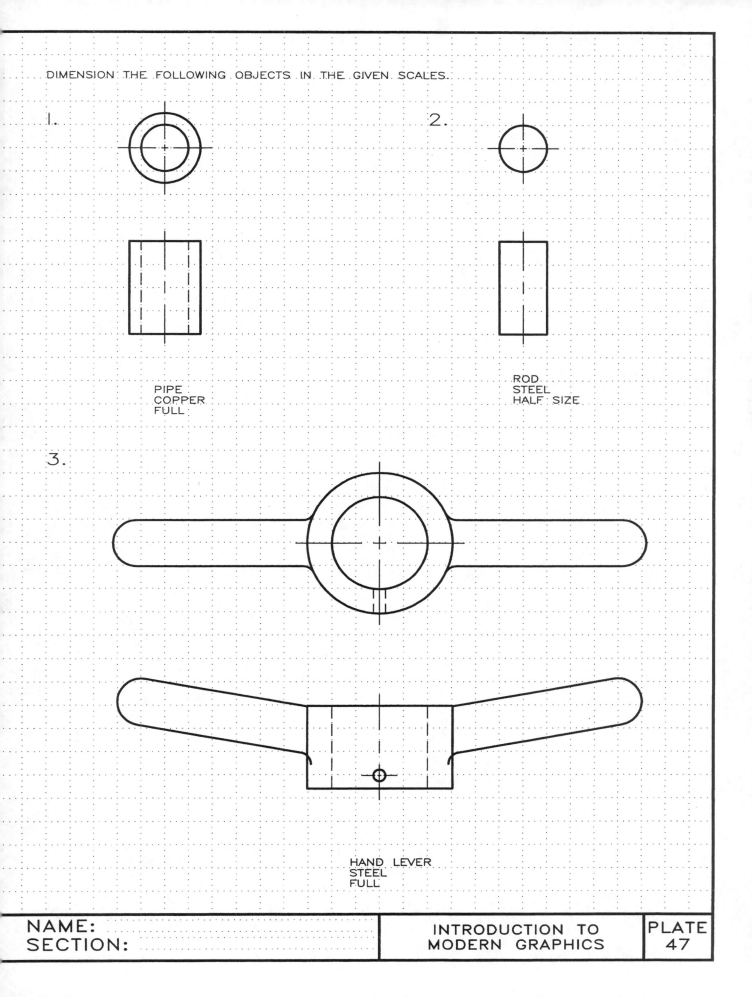

DIMENSION THE FOLLOWING OBJECTS IN THE GIVEN SCALES.

1.

2.

PIPE
COPPER
FULL

ROD
STEEL
HALF SIZE

3.

HAND LEVER
STEEL
FULL

NAME:
SECTION:

INTRODUCTION TO
MODERN GRAPHICS

PLATE
47

DIMENSION THE FOLLOWING MACHINED HOLES.
NOTE: THE THREADED HOLE IS COARSE SERIES.

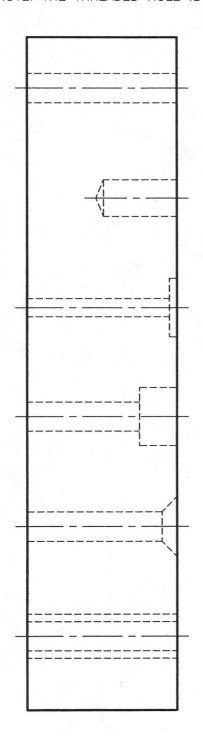

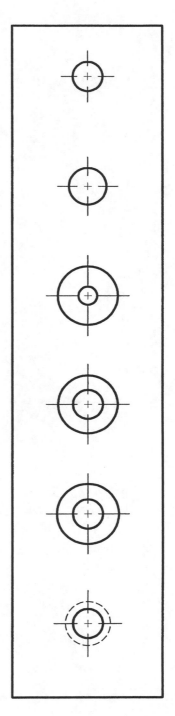

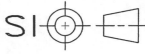

1. DIMENSION THE MOUNTING BRACKET. INCLUDE MACHINED HOLE NOTES AND A FILLET AND ROUND NOTE. ALL R=10. SCALE = 1:1 METRIC.

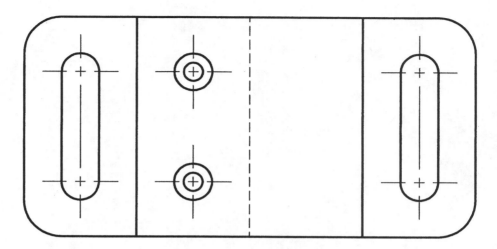

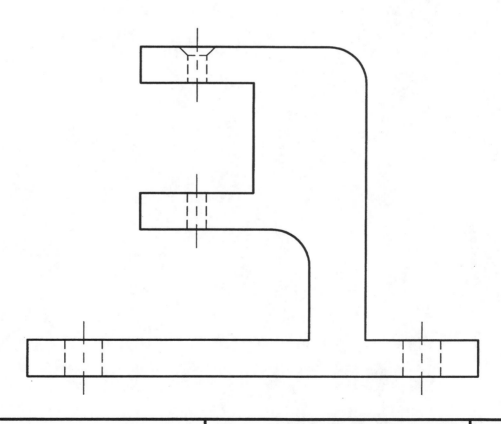

1.   DIMENSION THE MACHINED FEATURES OF THE WATCH STEM BELOW.   INCLUDE NOTES
FOR THE KNURL (PITCH=0.06), THE NECK AND THE POINT (1:5).

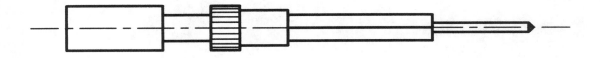

2.   DIMENSION THE MACHINED FEATURES OF THE JEWELRY SCREW DRIVER BELOW, INCLUDE
NOTES FOR THE KNURLS (96 DP), THE NECKS, CHAMFER, CONICAL TAPERS (1:3 AND 1:4
RATIO) AND THE STRAIGHT TAPER AT THE POINT (8° BOTH EDGES).

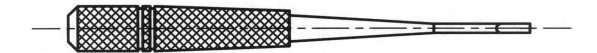

COMPLETELY DIMENSION THE DRAWING OF THE GUARD PLATE.
USE A NOTE TO INDICATE THE UNIFORM THICKNESS OF THIS PART.

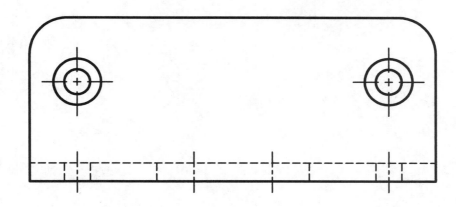

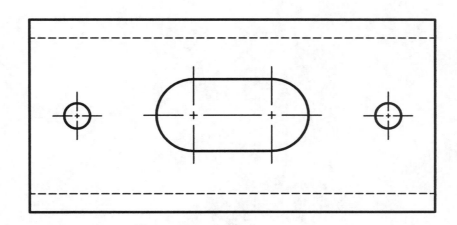

 SI

DIMENSION THE FOLLOWING OBJECTS. MEASURE THE DRAWINGS TO DETERMINE THE BASIC
SIZE AND CALCULATE THE LIMITS FOR EACH FIT.

1. RC1

2. LN2

3. LT4

4. FN1

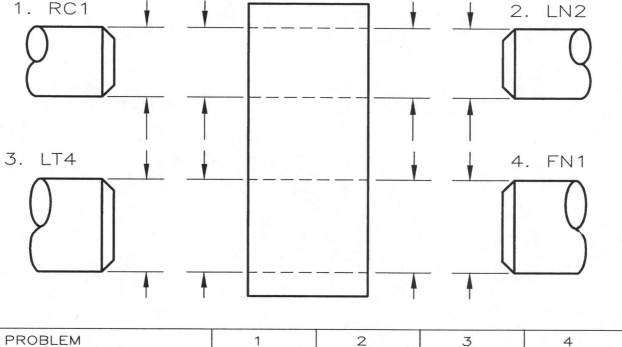

PROBLEM	1	2	3	4
ALLOWANCE				
MAX. CLEARANCE				
TYPE OF FIT				

5. H7/k6

6. H7/f7

7. P7/h6

8. U7/h6

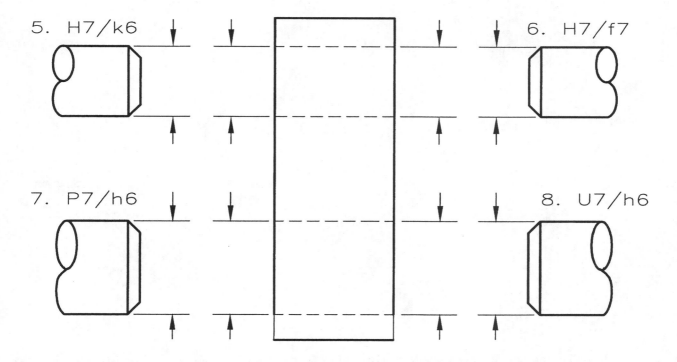

1. LABEL THE ELEMENTS OF THE FEATURE CONTORL FRAME.

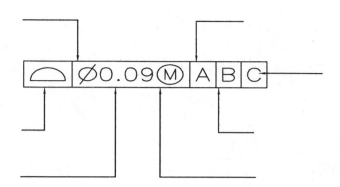

2. DIMENSION THE FOLLWING OBJECTS TO INDICATE THE REQUIRED TOLERANCES.

CYLINDRICITY 0.01
DIAMETER SIZE ± 0.02

PROFILE TOLERANCE 0.3
TOP SURFACE

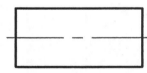

AXIS STRIAIGHTNESS 0.04
DIAMETER SIZE ±0.01

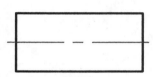

FLATNESS 0.06
TOP SURFACE

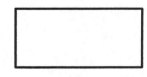

STRIAIGHTNESS 0.08
BOTTOM SURFACE

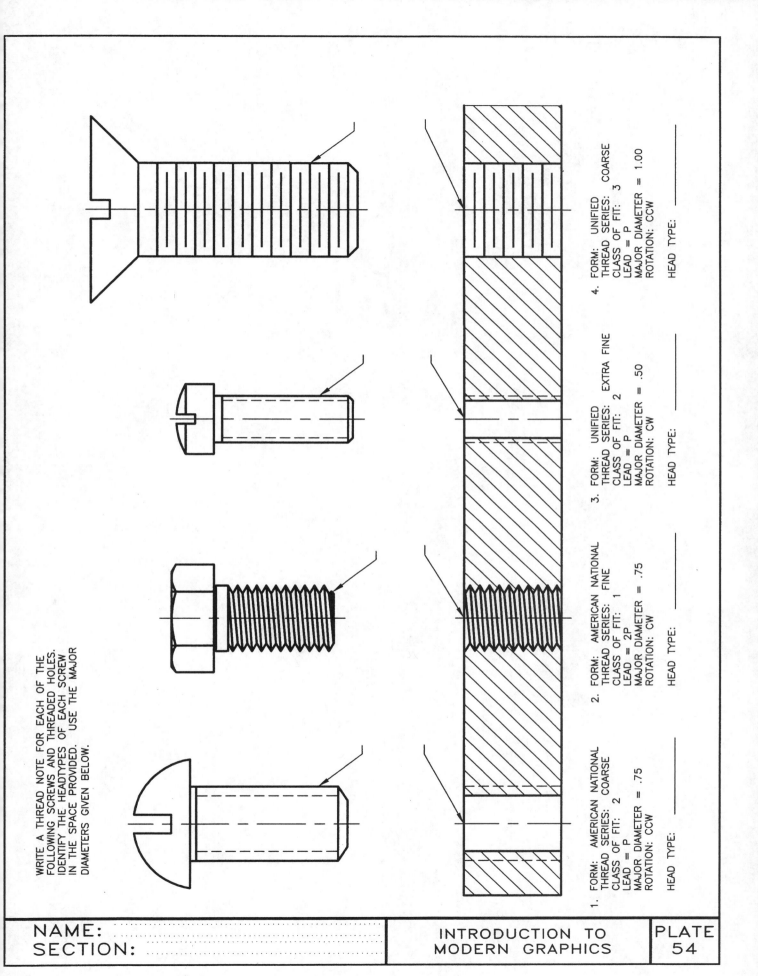

WRITE A THREAD NOTE FOR EACH OF THE
FOLLOWING SCREWS AND THREADED HOLES.
IDENTIFY THE HEADTYPES OF EACH SCREW
IN THE SPACE PROVIDED. USE THE MAJOR
DIAMETERS GIVEN BELOW.

1. FORM: AMERICAN NATIONAL
   THREAD SERIES: COARSE
   CLASS OF FIT: 2
   LEAD = P
   MAJOR DIAMETER = .75
   ROTATION: CCW

   HEAD TYPE: _____

2. FORM: AMERICAN NATIONAL
   THREAD SERIES: FINE
   CLASS OF FIT: 1
   LEAD = 2P
   MAJOR DIAMETER = .75
   ROTATION: CW

   HEAD TYPE: _____

3. FORM: UNIFIED
   THREAD SERIES: EXTRA FINE
   CLASS OF FIT: 2
   LEAD = P
   MAJOR DIAMETER = .50
   ROTATION: CW

   HEAD TYPE: _____

4. FORM: UNIFIED
   THREAD SERIES: COARSE
   CLASS OF FIT: 3
   LEAD = P
   MAJOR DIAMETER = 1.00
   ROTATION: CCW

   HEAD TYPE: _____

NAME: ....
SECTION: ....

INTRODUCTION TO
MODERN GRAPHICS

PLATE
54

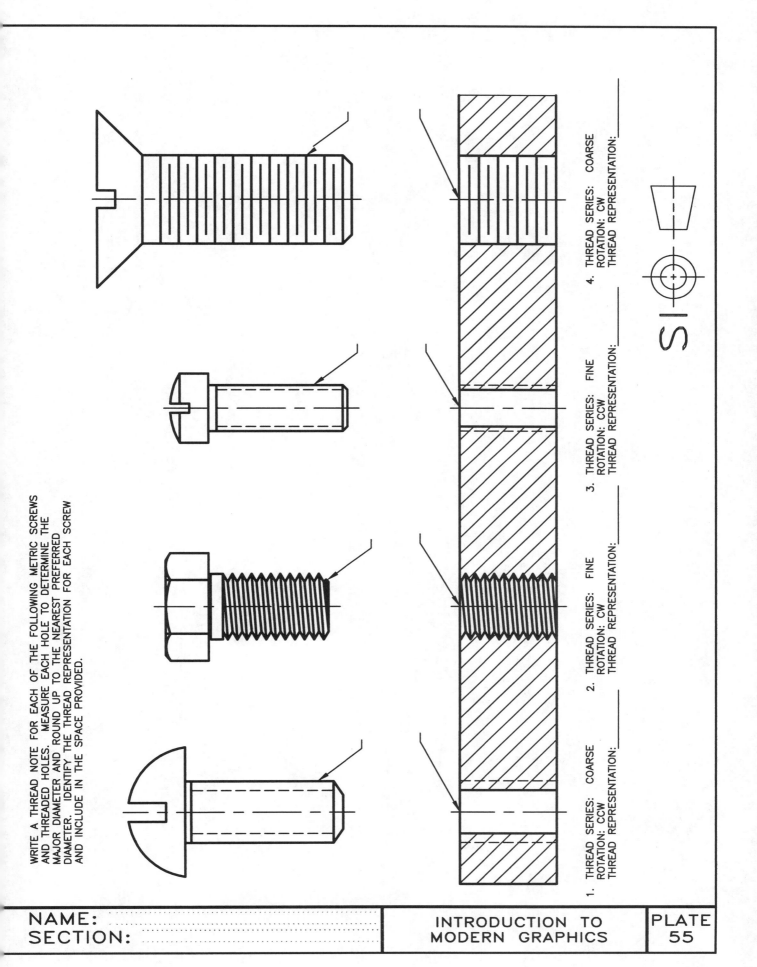

WRITE A THREAD NOTE FOR EACH OF THE FOLLOWING METRIC SCREWS AND THREADED HOLES. MEASURE EACH HOLE TO DETERMINE THE MAJOR DIAMETER AND ROUND UP TO THE NEAREST PREFERRED DIAMETER. IDENTIFY THE THREAD REPRESENTATION FOR EACH SCREW AND INCLUDE IN THE SPACE PROVIDED.

1. THREAD SERIES: COARSE
   ROTATION: CCW
   THREAD REPRESENTATION: _____

2. THREAD SERIES: FINE
   ROTATION: CW
   THREAD REPRESENTATION: _____

3. THREAD SERIES: FINE
   ROTATION: CCW
   THREAD REPRESENTATION: _____

4. THREAD SERIES: COARSE
   ROTATION: CW
   THREAD REPRESENTATION: _____

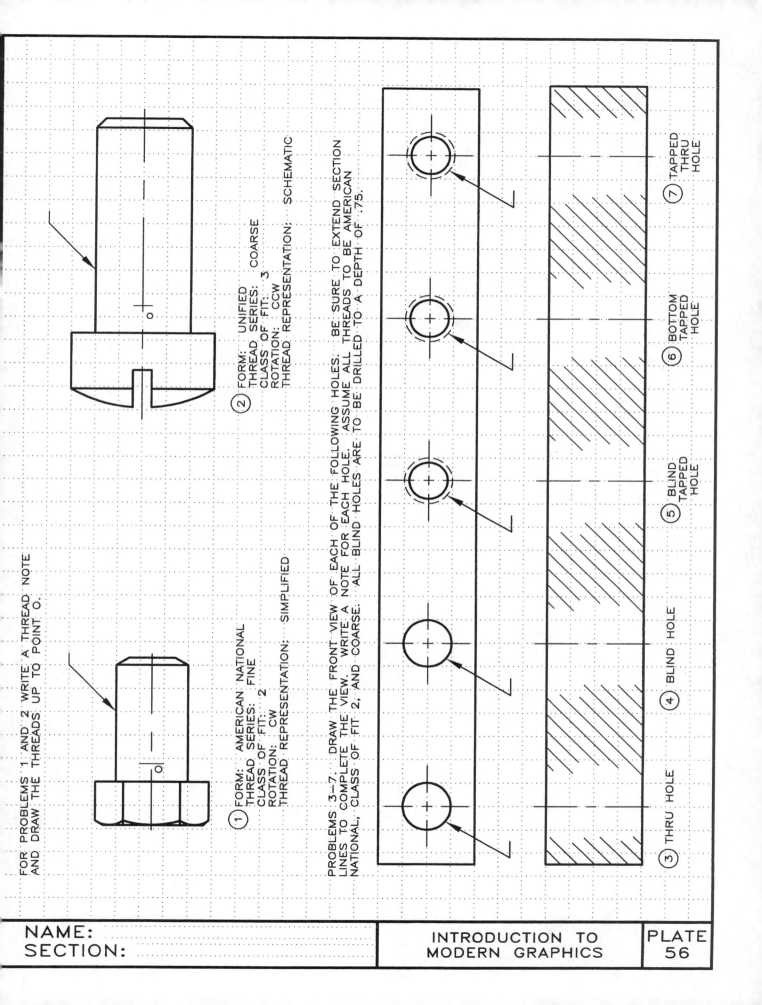

FOR PROBLEMS 1 AND 2 WRITE A THREAD NOTE
AND DRAW THE THREADS UP TO POINT O.

(1) FORM: AMERICAN NATIONAL
    THREAD SERIES: FINE
    CLASS OF FIT: 2
    ROTATION: CW
    THREAD REPRESENTATION: SIMPLIFIED

(2) FORM: UNIFIED
    THREAD SERIES: COARSE
    CLASS OF FIT: 3
    ROTATION: CCW
    THREAD REPRESENTATION: SCHEMATIC

PROBLEMS 3-7. DRAW THE FRONT VIEW OF EACH OF THE FOLLOWING HOLES. BE SURE TO EXTEND SECTION
LINES TO COMPLETE THE VIEW. WRITE A NOTE FOR EACH HOLE. ASSUME ALL THREADS TO BE AMERICAN
NATIONAL, CLASS OF FIT 2, AND COARSE. ALL BLIND HOLES ARE TO BE DRILLED TO A DEPTH OF .75.

(3) THRU HOLE   (4) BLIND HOLE   (5) BLIND TAPPED HOLE   (6) BOTTOM TAPPED HOLE   (7) TAPPED THRU HOLE

NAME:
SECTION:

INTRODUCTION TO
MODERN GRAPHICS

PLATE
56

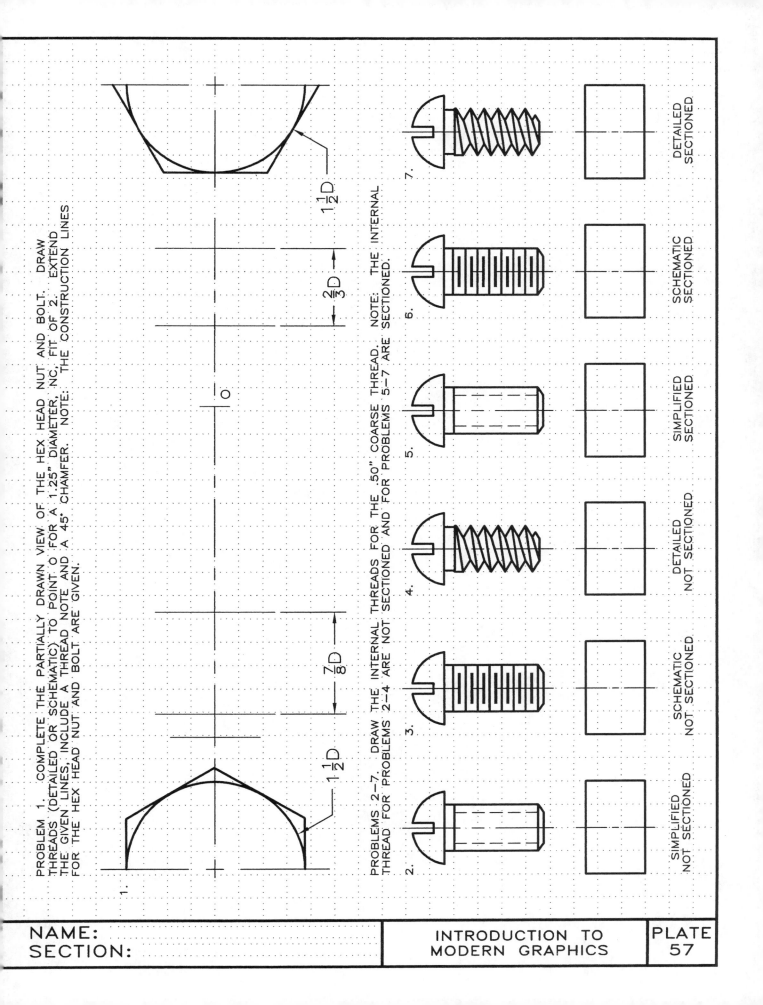

PROBLEM 1. COMPLETE THE PARTIALLY DRAWN VIEW OF THE HEX HEAD NUT AND BOLT. DRAW THREADS (DETAILED OR SCHEMATIC) TO POINT O FOR A 1.25" DIAMETER, NC, FIT OF 2. EXTEND THE GIVEN LINES, INCLUDE A THREAD NOTE AND A 45° CHAMFER. NOTE: THE CONSTRUCTION LINES FOR THE HEX HEAD NUT AND BOLT ARE GIVEN.

1.

$1\frac{1}{2}$D

$\frac{7}{8}$D

O

$1\frac{1}{2}$D

$\frac{2}{3}$D

PROBLEMS 2-7. DRAW THE INTERNAL THREADS FOR THE .50" COARSE THREAD. NOTE: THE INTERNAL THREAD FOR PROBLEMS 2-4 ARE NOT SECTIONED AND FOR PROBLEMS 5-7 ARE SECTIONED.

2. SIMPLIFIED NOT SECTIONED

3. SCHEMATIC NOT SECTIONED

4. DETAILED NOT SECTIONED

5. SIMPLIFIED SECTIONED

6. SCHEMATIC SECTIONED

7. DETAILED SECTIONED

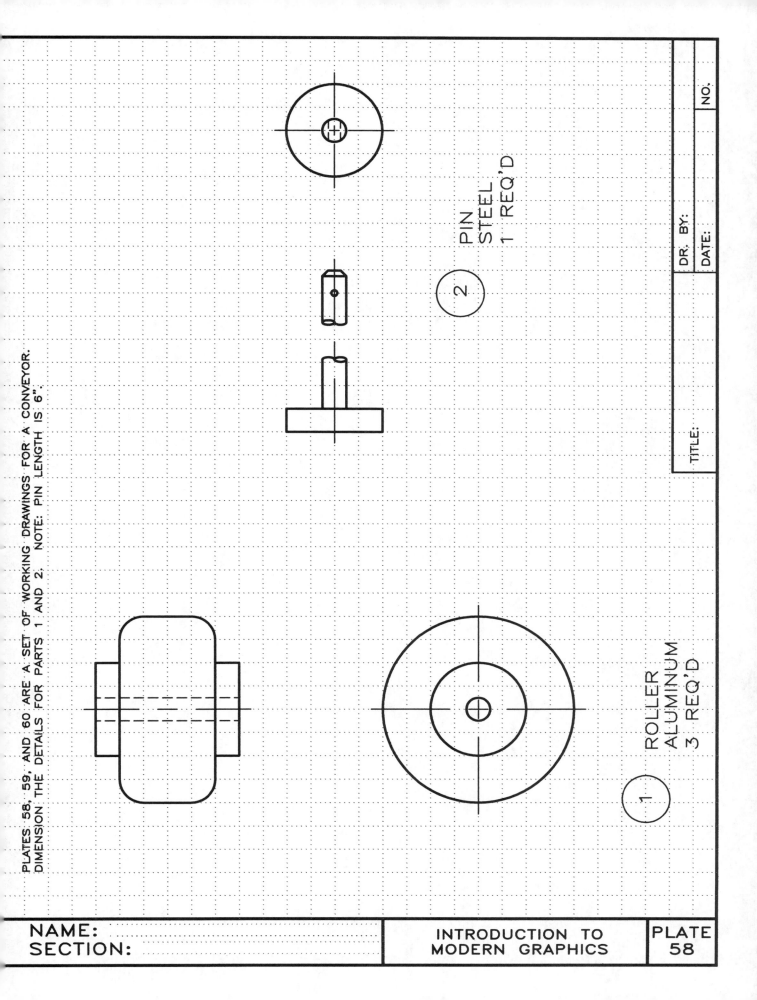

PLATES 58, 59, AND 60 ARE A SET OF WORKING DRAWINGS FOR A CONVEYOR.
DIMENSION THE DETAILS FOR PARTS 1 AND 2. NOTE: PIN LENGTH IS 6".

PIN
STEEL
1 REQ'D

2

ROLLER
ALUMINUM
3 REQ'D

1

NAME:
SECTION:

INTRODUCTION TO
MODERN GRAPHICS

PLATE
58

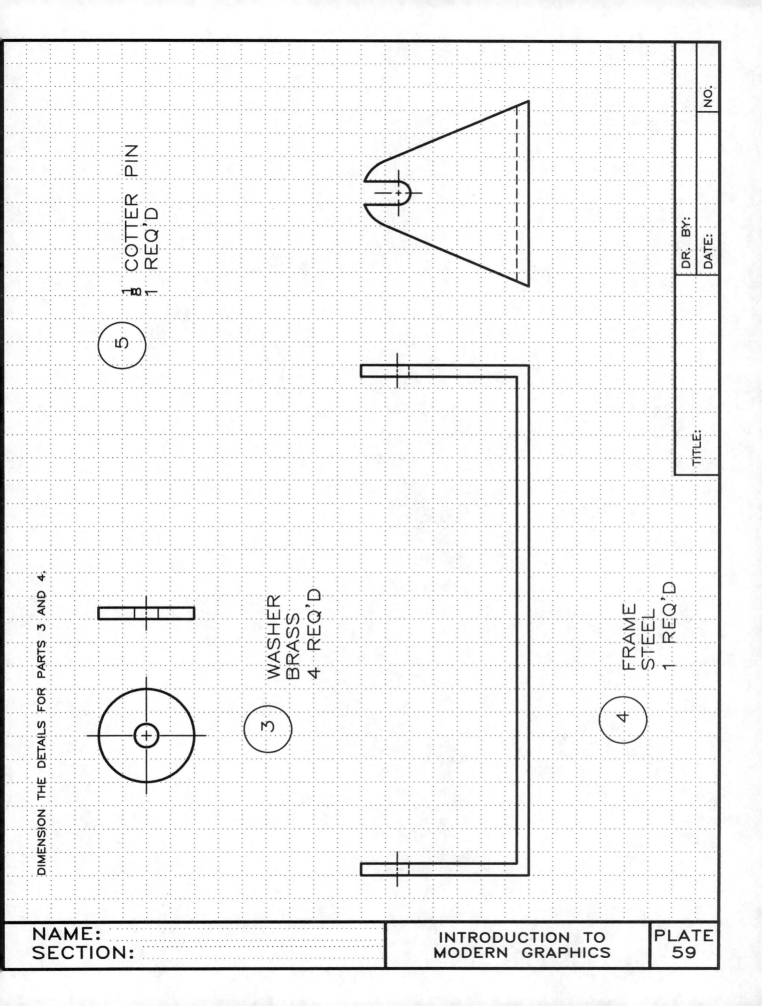

DIMENSION THE DETAILS FOR PARTS 3 AND 4.

⑤ COTTER PIN
    $\frac{1}{8}$
    1 REQ'D

③ WASHER
    BRASS
    4 REQ'D

④ FRAME
    STEEL
    1 REQ'D

NAME:
SECTION:

INTRODUCTION TO
MODERN GRAPHICS

PLATE
59

DR. BY:
DATE:
TITLE:
NO.

NO.					PART NAME		REQ'D	MATERIAL		

DR. BY:

DATE:

TITLE:

NO.

NAME:

SECTION:

INTRODUCTION TO
MODERN GRAPHICS

PLATE
60

DESIGN A PIPE CLAMP SIMILAR TO THE ONE SHOWN BELOW. THE CLAMP NEEDS
TO ACCOMODATE A 44mm PIPE. DRAW THE DETAILS FOR EACH PART. BE SURE
TO INCLUDE DIMENSIONS AND PART LABELS.

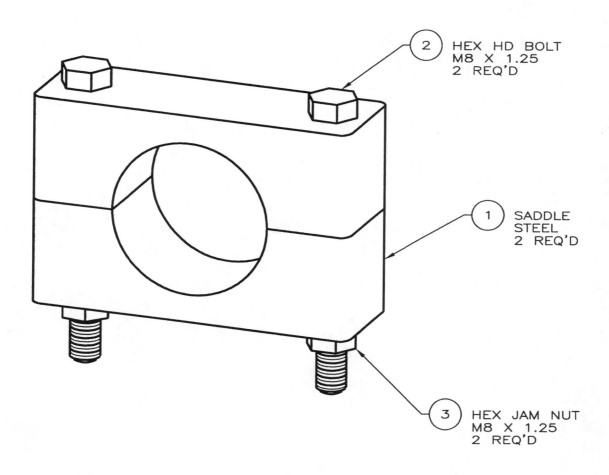

2 HEX HD BOLT
M8 X 1.25
2 REQ'D

1 SADDLE
STEEL
2 REQ'D

3 HEX JAM NUT
M8 X 1.25
2 REQ'D

SI

DRAW AN ASSEMBLY WITHOUT HIDDEN LINES FOR THE PIPE CLAMP DESIGNED ON
PLATE 61.   BE SURE TO INCLUDE BALLOON NUMBERS AND COMPLETE THE PARTS
LIST.

SI

NO.	PART NAME	REQ'D	MATERIAL

NAME:...

SECTION:...

INTRODUCTION TO
MODERN GRAPHICS

PLATE
62

DRAW A SECTIONED ASSEMBLY FOR THE PIPE CLAMP DESIGNED ON PLATE 61.   BE
SURE TO INCLUDE BALLOON NUMBERS AND COMPLETE THE PARTS LIST.

SI

NO.	PART NAME	REQ'D	MATERIAL

NAME:

SECTION:

INTRODUCTION TO
MODERN GRAPHICS

PLATE
63

DRAW AN EXPLODED ASSEMBLY FOR THE PIPE CLAMP DESIGNED ON PLATE 61.
BE SURE TO INCLUDE BALLOON NUMBERS AND COMPLETE THE PARTS LIST.

NO.	PART NAME	REQ'D	MATERIAL

SI

NAME:
SECTION:

INTRODUCTION TO
MODERN GRAPHICS

PLATE
64

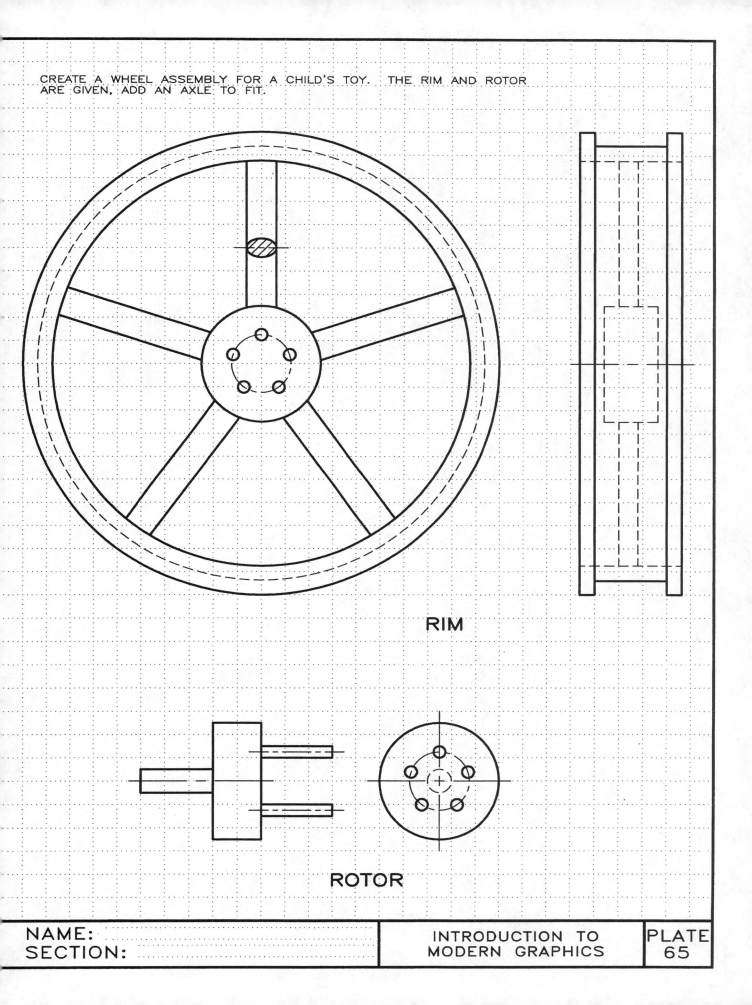

CREATE A WHEEL ASSEMBLY FOR A CHILD'S TOY. THE RIM AND ROTOR
ARE GIVEN, ADD AN AXLE TO FIT.

RIM

ROTOR

SKETCH A SET OF WORKING DRAWINGS FOR THE HANDWHEEL SHOWN.

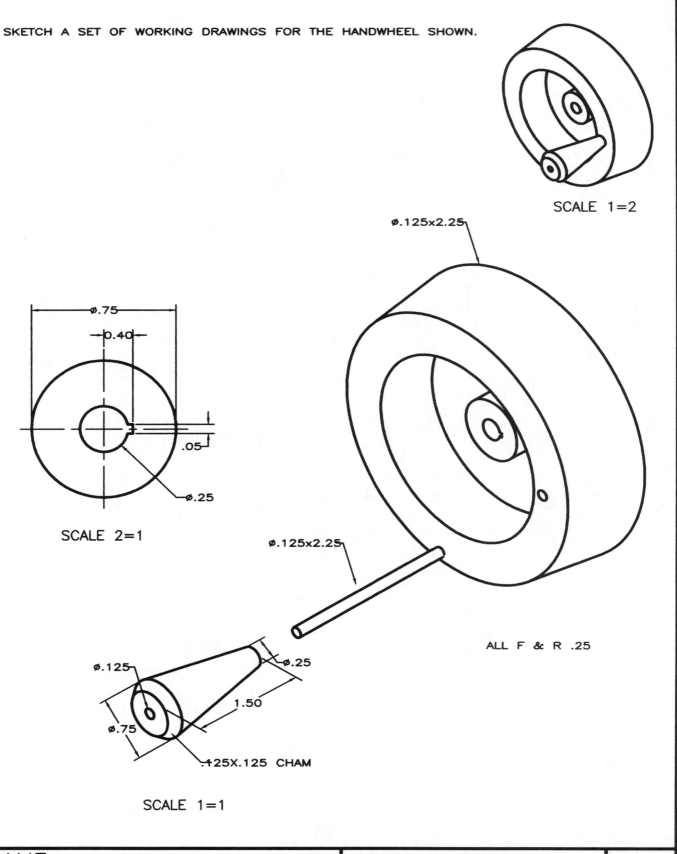

SCALE 1=2

ø.125x2.25

ø.75

Ø.40

.05

Ø.25

SCALE 2=1

ø.125x2.25

ALL F & R .25

Ø.25

ø.125

1.50

Ø.75

.125X.125 CHAM

SCALE 1=1

CREATE A SET OF WORKING DRAWINGS FOR A DOOR KNOCKER. THE PICTURE BELOW IS AN EXAMPLE OBJECT.   CREATE YOUR OWN DESIGN.

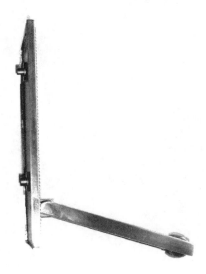

PLAN YOUR DESIGN BY MAKING A ROUGH SKETCH NOT TO SCALE (NTS).

CONSIDER:

WHAT SIZE SHOULD YOU MAKE THE FINISHED PART?
HOW MANY PARTS WILL YOU NEED?
WHICH PARTS WILL BE STANDARD PARTS?
HOW WILL YOU ATTACH THE RING TO THE BASE PLATE?
HOW WILL YOU ATTACH THE DOOR KNOCKER TO THE DOOR?

YOUR DRAWING SHOULD INCLUDE
- o   TWO VIEWS FOR EACH PART
- o   A LABEL EACH PART
- o   A PARTS LIST FOR YOUR DESIGN

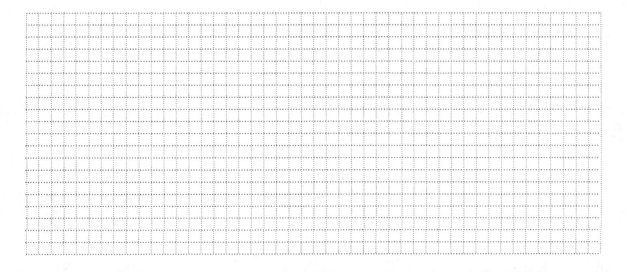

NAME:

SECTION:

INTRODUCTION TO
MODERN GRAPHICS

PLATE
67

CONSTRUCT ISOMETRIC DRAWINGS OF EACH OF THE OBJECTS SHOWN BELOW.

1.

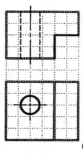

O

2.

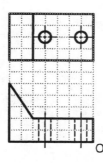

O

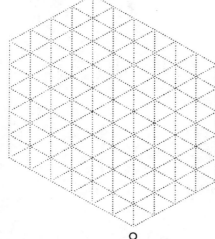

O

O

3.

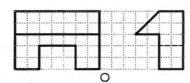

O

4.

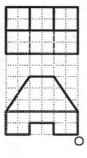

O

O

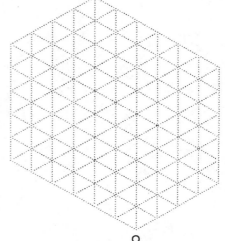

O

CONSTRUCT ISOMETRIC DRAWINGS OF EACH OF THE OBJECTS SHOWN BELOW.

1.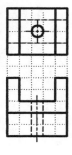

O

2.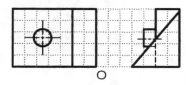

O

O

3.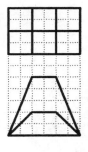

O

4.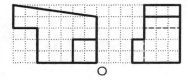

O

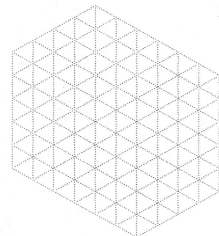

O

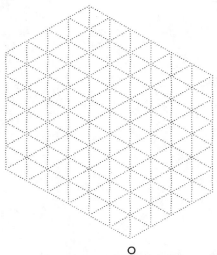

O

O

CONSTRUCT AN ISOMETRIC FOR EACH OF THE PROBLEMS SHOWN BELOW.

1.

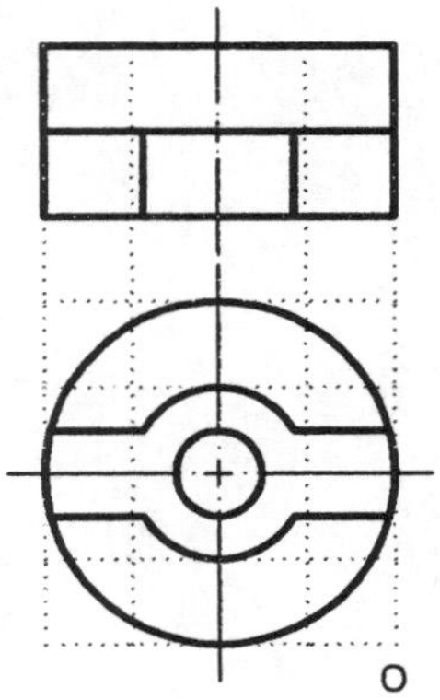

2.

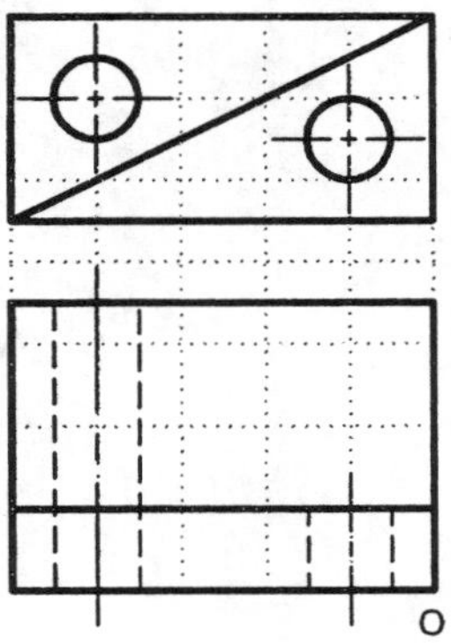

3.

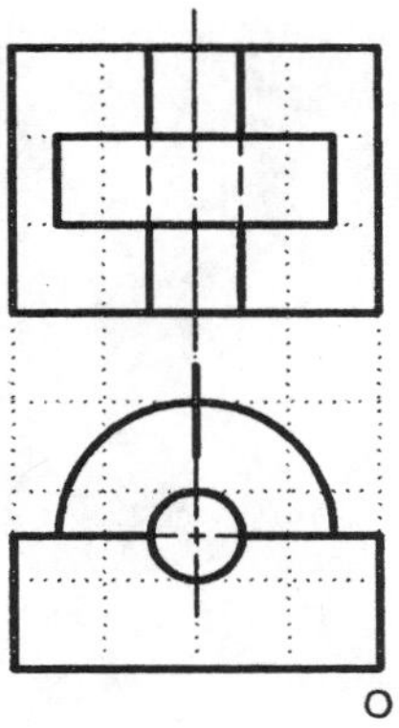

4.

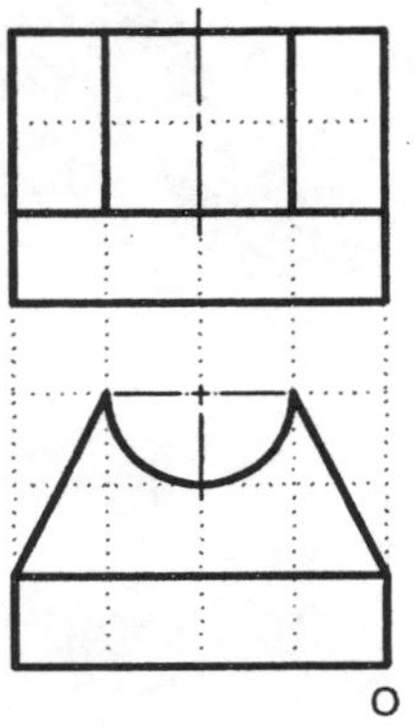

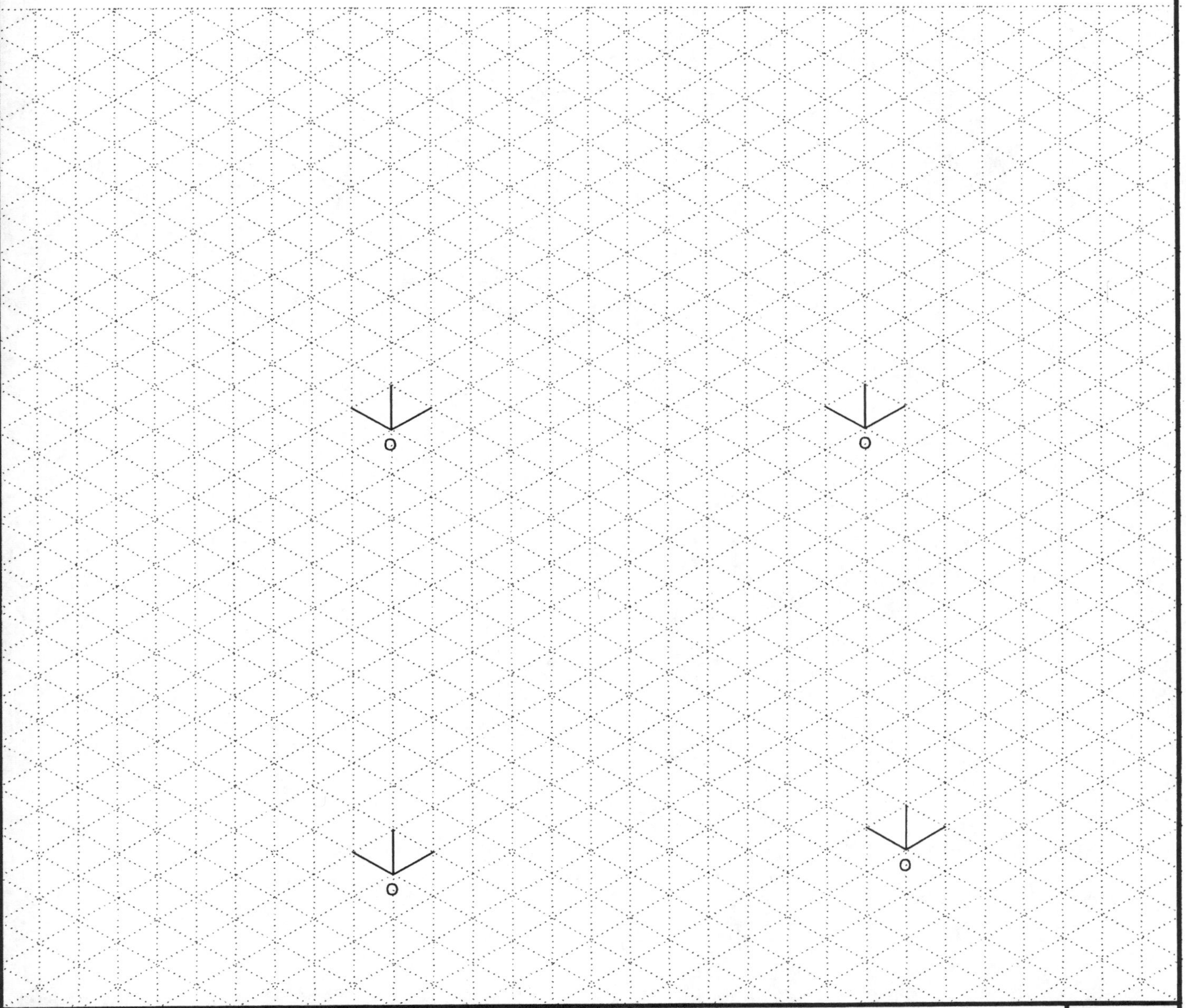

O

O

O

O

O

O

O

O

COMPLETE ISOMETRIC SKETCHES
FOR EACH OF THE OBJECTS
SHOWN.

1.

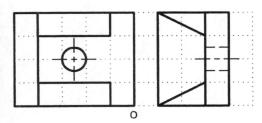

2.

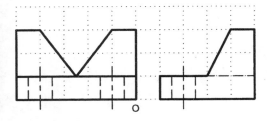

3.

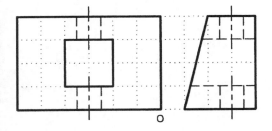

4.

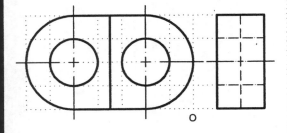

NAME:

SECTION:

INTRODUCTION TO
MODERN GRAPHICS

PLATE
71

SKETCH AN ISOMETRIC OF THE FLAG POLE MOUNT ON THE GRID PROVIDED.

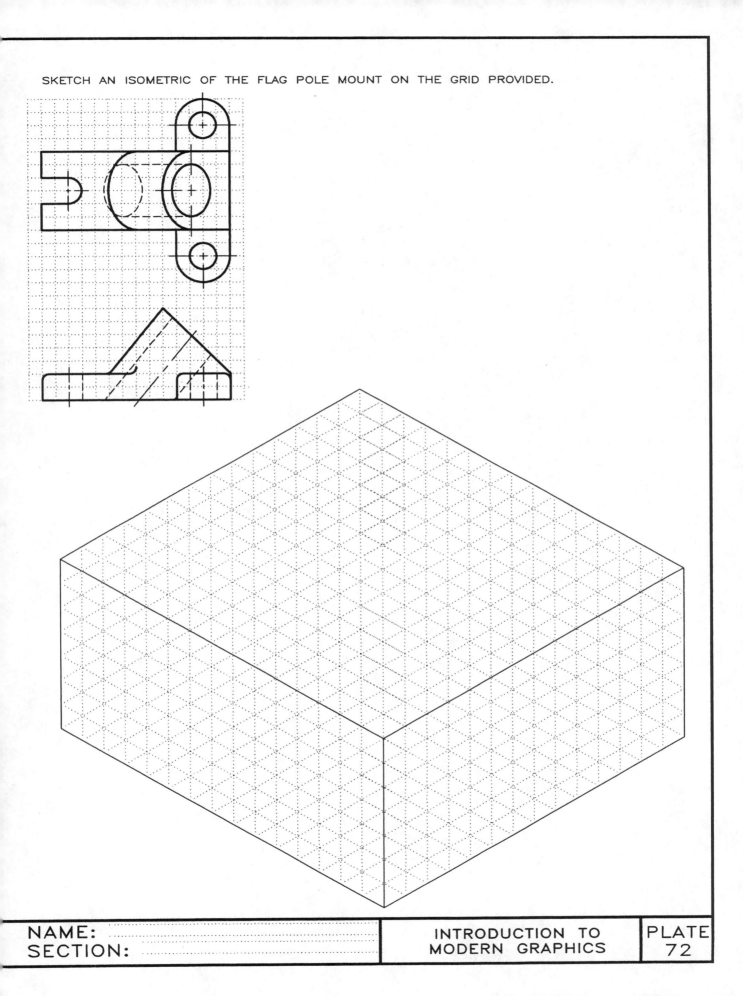

DRAW CAVALIER OBLIQUES OF THE FOLLOWING OBJECTS.   USE THE
INDICATED RECEDING ANGLES.

1. U—BLOCK

2. BRACE BLOCK

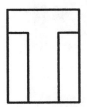

30°

45°

3. A—FRAME BLOCK

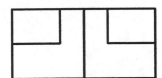

4. PIPE CRADLE

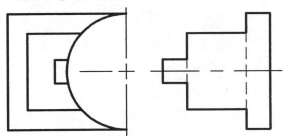

60°

45°

NAME: ....
SECTION: ....

INTRODUCTION TO
MODERN GRAPHICS

PLATE
73

DRAW THE SHOE RACK AS
AS A CABINET AND CAVALIER OBLIQUE.
USE A RECEDING ANGLE OF 45°

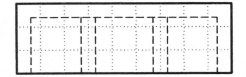

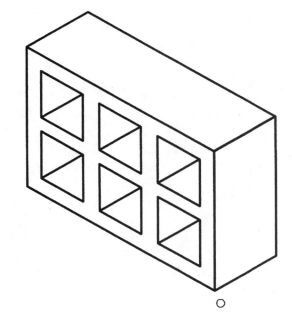

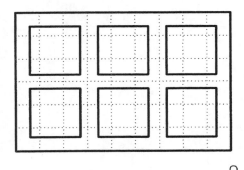

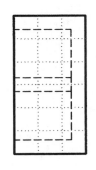

SHOE RACK
SCALE 1/2"=1'-0

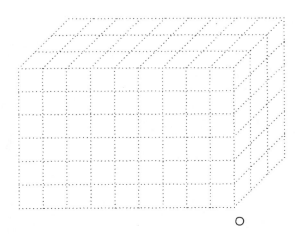

CAVALIER

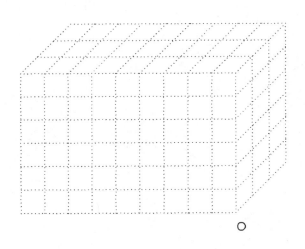

CABINET

DRAW THE BUCKLE AS A
CABINET OBLIQUE. USE A RECEDING
ANGLE OF 45 DEGREES.

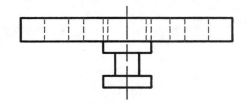

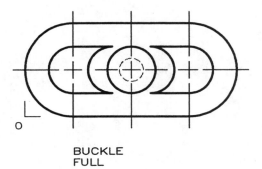

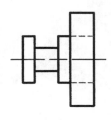

O

BUCKLE
FULL

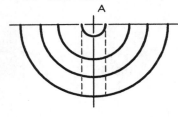

O

DRAW THE KNOB AS A CAVALIER OBLIQUE. USE A RECEDING ANGLE OF
30 DEGREES.

A

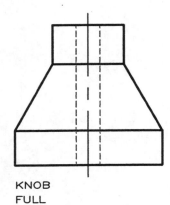

$\overset{+}{A}$

KNOB
FULL

SKETCH THE OBLIQUE REPRESENTATION OF THE FOLLOWING OBJECTS. DRAW A CABINET
OR CAVILER OBLIQUE AS ASSIGNED.

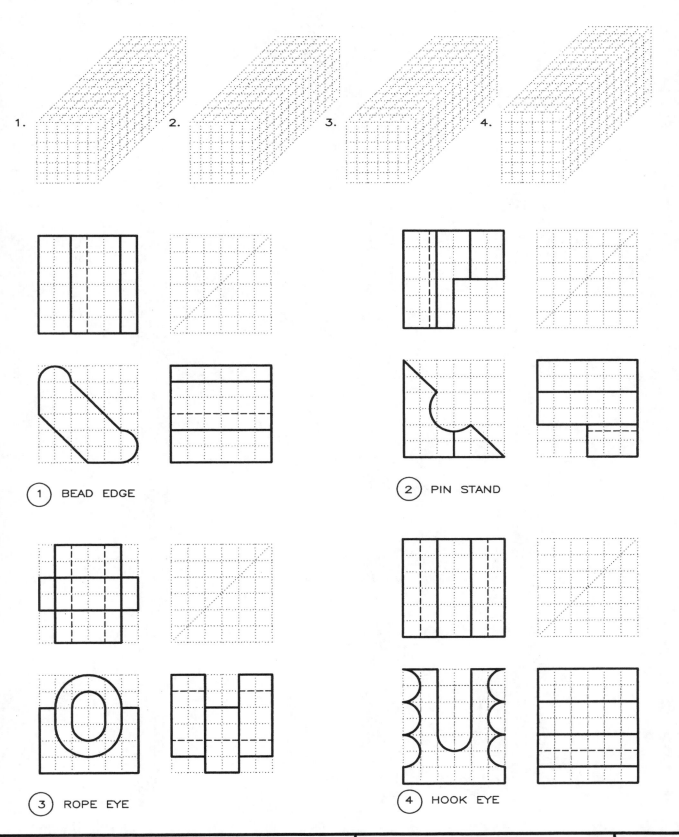

1.

2.

3.

4.

① BEAD EDGE

② PIN STAND

③ ROPE EYE

④ HOOK EYE

DRAW A CAVALIER OBLIQUE OF THE OBJECT BELOW.

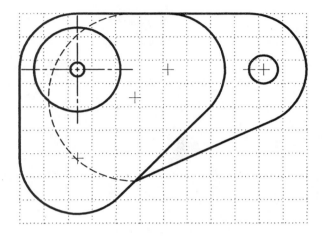 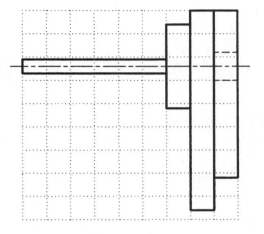

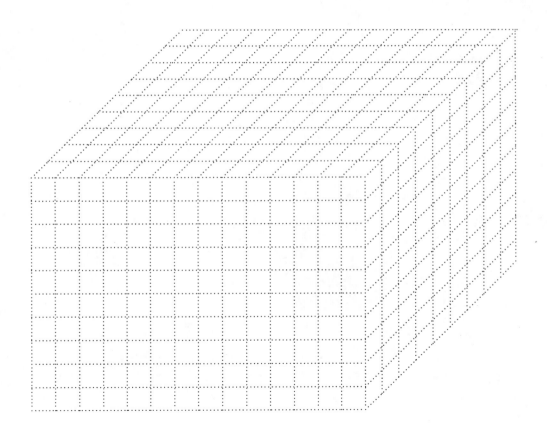

DRAW A ONE—POINT PERSPECTIVE OF THE SOFA SHOWN BELOW USING THE GIVEN
VANISHING POINT. ESTIMATE THE DEPTH TO GIVE A REALISTIC REPRESENTATION OF THE
SOFA.

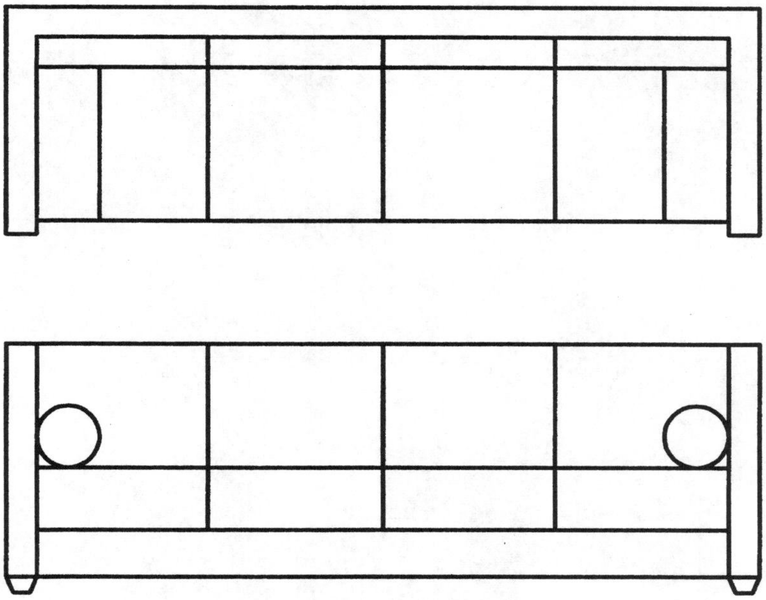

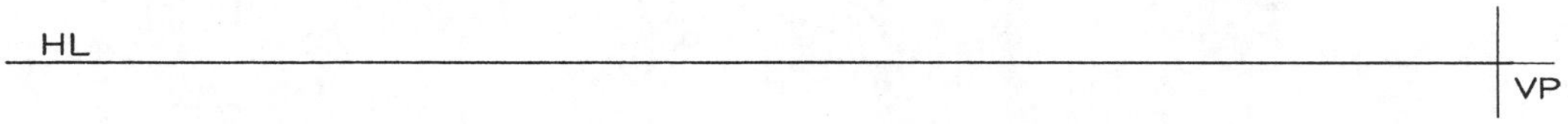

HL

VP

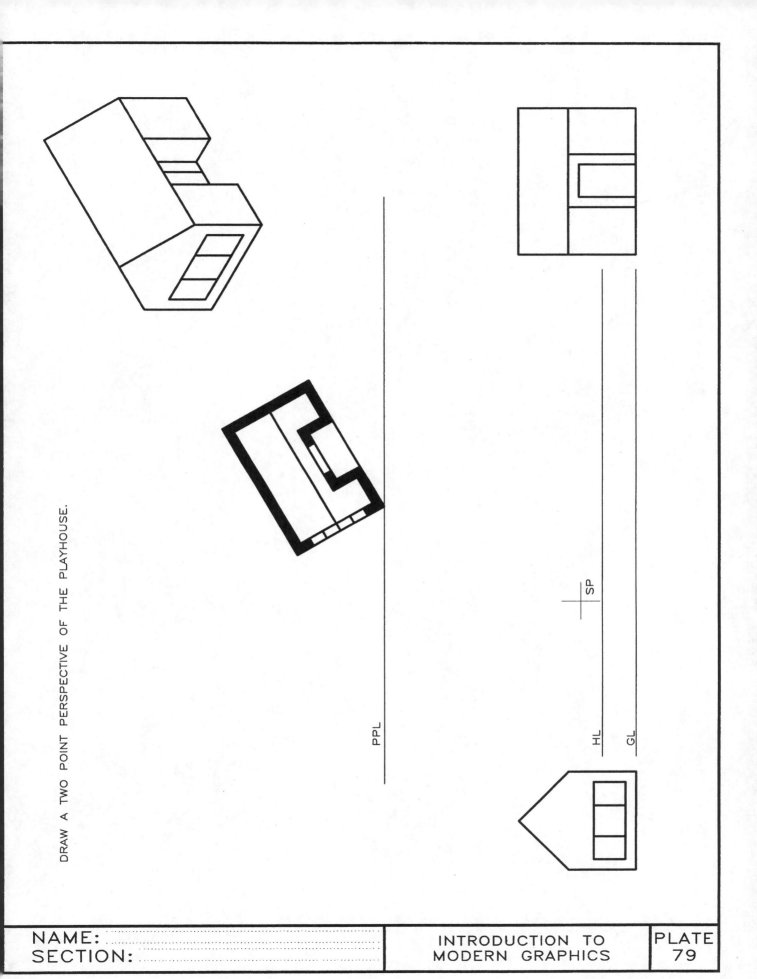

DRAW A TWO POINT PERSPECTIVE OF THE PLAYHOUSE.

PPL

SP

HL

GL

DESIGN A PLATE CAM TO GENERATE THE FOLLOWING MOTION IN ONE REVOLUTION.
HARMONIC RISE FROM 0 TO 150 DEGREES
DWELL FROM 150 TO 210 DEGREES
HARMONIC FALL FROM 210 TO 360 DEGREES

THE BASE CIRCLE DIAMETER IS 1.50 INCHES AND A DISPLACEMENT OF 1.50
INCHES.   THE ROTATION IS CLOCKWISE.

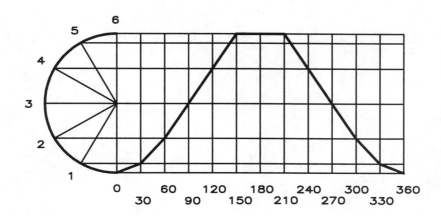

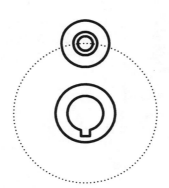

DESIGN A PLATE CAM TO GENERATE THE FOLLOWING MOTION IN ONE REVOLUTION.

DWELL FROM 0 TO 60 DEGREES
HAROMONIC RISE FROM 60 TO 210 DEGREES
hARMONIC FALL FROM 210 TO 300 DEGREES
DWELL FROM 300 TO 330 DEGREES
STRAIGHT FALL FROM 330 TO 360

TOTAL DISPLACEMENT IS 1.25 INCHES.   THE BASE CIRCLE DIAMETER IS 1.50
INCHES.   THE ROTATION IS COUNTER CLOCKWISE.

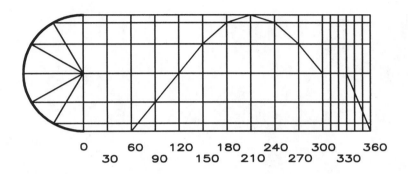

0   30   60   90   120   150   180   210   240   270   300   330   360

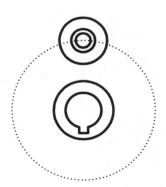

INTRODUCTION TO
MODERN GRAPHICS

PLATE
81

REVISE THE GIVEN CIRCUIT DRAWINGS USING THE CORRECT SYMBOLS FOR EACH
OF THE INDICATED COMPONENTS TO CREATE THE SCHEMATIC.   REPLACE ALL
TEXT DESCRIPTIONS WITH THE CORRECT COMPONENT SYMBOL FOR ALL TEXT
EXCEPT FOR THE TEXT ENCLOSED WITHIN BOXES. ALL WIRES ARE CONNECTING.

1. A LAYOUT FOR A CIRCUIT THAT CAUSES THE CAR LIGHTS TO REMAIN ON FOR
A FEW SECONDS AFTER THE DOOR IS CLOSED.

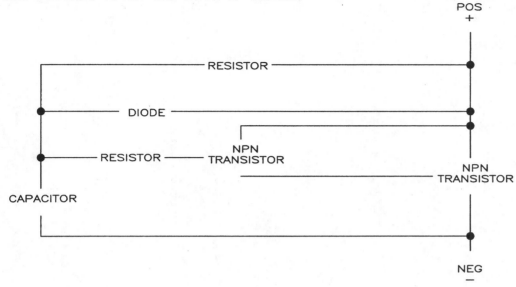

2. A LAYOUT FOR A CIRCUIT THAT DELAYS THE RINGING OF CHIME IF THE
BUTTON IS PRESSED MULTIPLE TIMES.

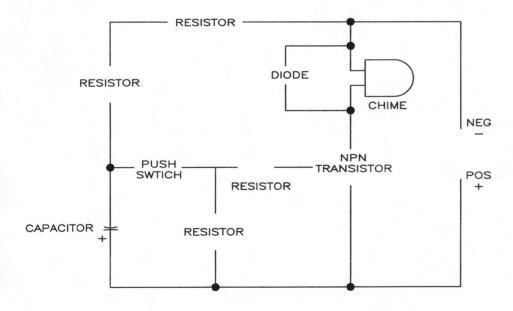

REVISE THE GIVEN CIRCUIT DRAWINGS USING THE CORRECT SYMBOLS FOR EACH
OF THE INDICATED COMPONENTS TO CREATE THE SCHEMATIC. REPLACE ALL
TEXT DESCRIPTIONS WITH THE CORRECT COMPONENT SYMBOL FOR ALL TEXT
EXCEPT FOR THE TEXT ENCLOSED WITHIN BOXES. ALL WIRES ARE CONNECTING.

1. A LAYOUT FOR A CIRCUIT THAT SOUNDS AN ALARM WHEN THE WATER LEVEL
IN CONTAINER DROPS.

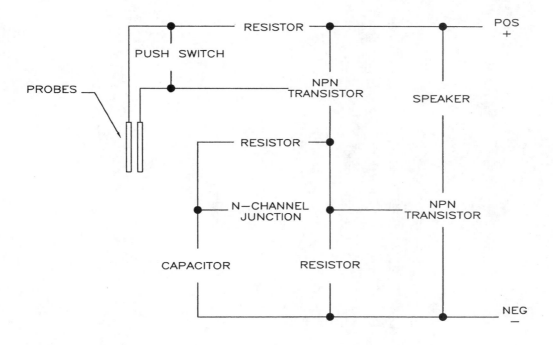

2. A CIRCUIT LAYOUT THAT ALLOWS AUTO WINDSHIELD WIPERS TO OPERATE AT
INTERVALS.

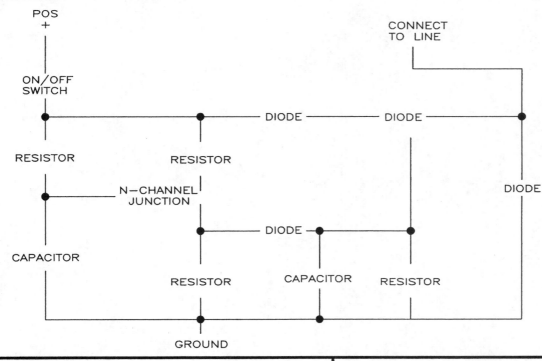

1. DRAW A PROFILE OF THE LANDFORM ALONG A STRAIGHT LINE FROM
MIDPOINTS A—A.   USE A SPLINE CURVE TO REFINE YOUR PLOT.

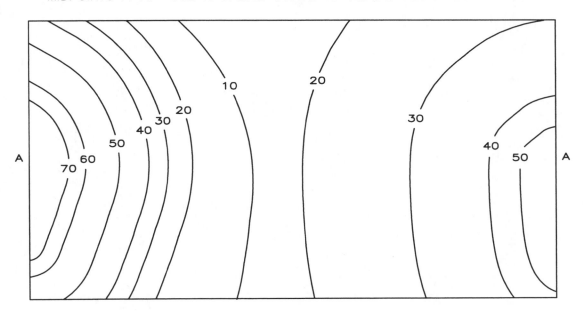

2. DRAW A PROFILE OF THE INDIAN MOUNDS ALONG A STRAIGHT LINE
THROUGH THE HIGHEST POINTS. USE A SPLINE CURVE TO REFINE YOUR
PLOT.

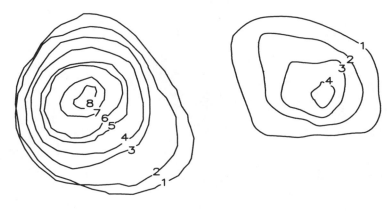

DRAW THE CUT AND FILL DIAGRAM FOR THE LEVEL ROAD IN THE FRONT VIEW
AND PLOT THE CONTOUR LINES.   ROAD ELEVATION IS 40 FEET; THE ROAD IS 12
FEET WIDE. CUT ANGLE IS 30 DEGREES; FILL ANGLE IS 45 DEGREES.

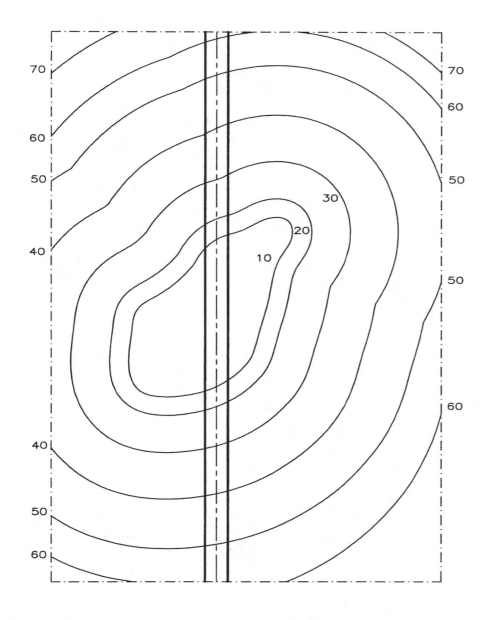

70
60
50
40
30
20
10

SCALE 1"=50'

DRAW THE FRONT VIEW OF THE PIPE LAYOUT. USE SINGLE-LINE SYMBOLS.

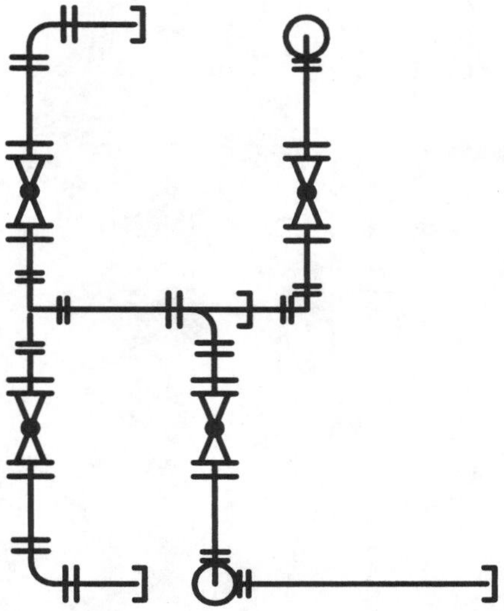

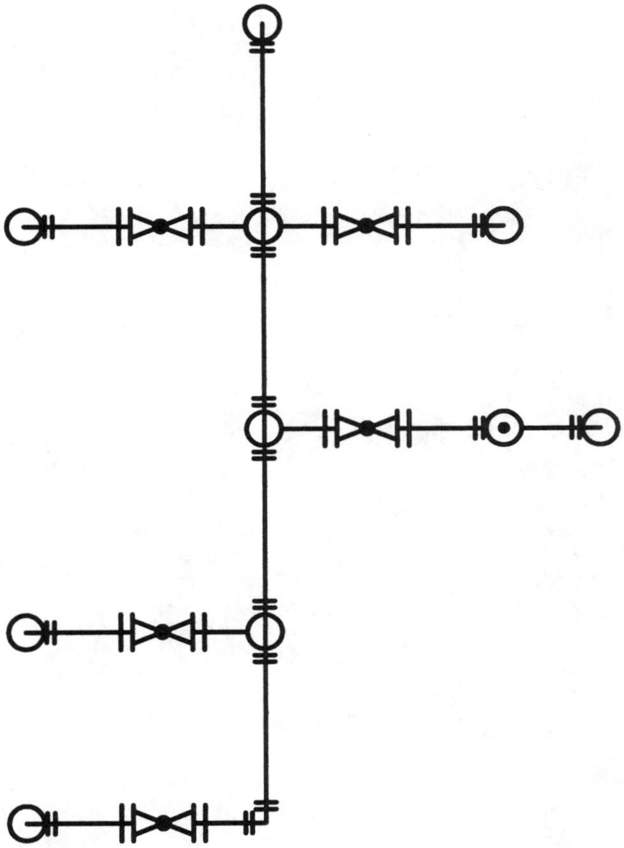

NAME:

SECTION:

INTRODUCTION TO MODERN GRAPHICS

PLATE 86

REDRAW THE PIPE LAYOUT USING SINGLE LINE SYMBOLS.

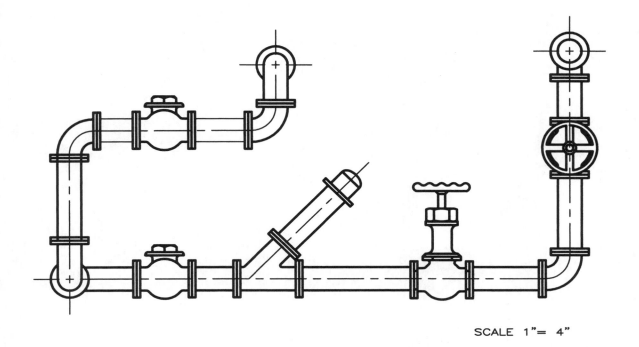

SCALE 1"= 4"

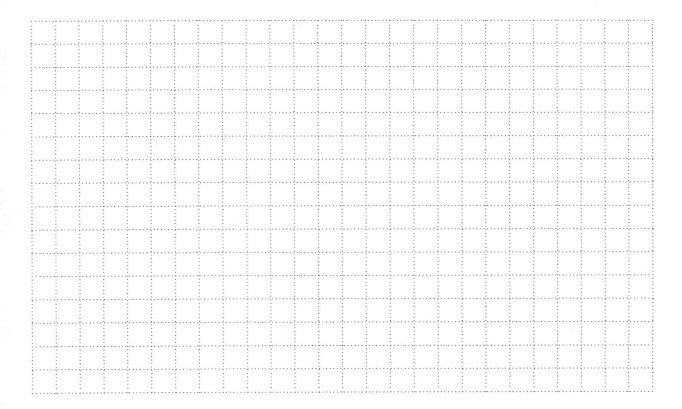

COMPLETE THE WELDING NOTATIONS FOR THE INDICATED WELDS.

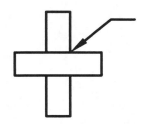

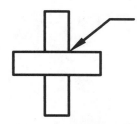

1. ADD A WELD NOTE FOR A 6 MM CONTINUOUS FILLET WELD ON THE TOP RIGHT JOINT OF THE BRACE.

2. ADD A WELD NOTE FOR A 6 MM CONTINUOUS FILLET WELD ON THE TOP LEFT JOINT OF THE BRACE.

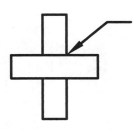

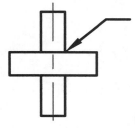

3. ADD A WELD NOTE FOR A 6 MM CONTINUOUS FILLET ON THE RIGHT AND LEFT TOP JOINTS OF THE BRACKET.

4. ADD A WELD NOTE FOR A 6 MM CONTINUOUS FILLET WELD THAT WILL BE APPLIED ON ALL JOINTS ON THE BOTTOM EDGE OF THE BRACKET.

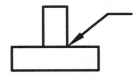

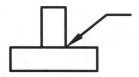

5 ADD A WELD NOTE FOR 6 MM INTERMITTENT FILLET WELDS ON THE RIGHT JOINT OF THE BRACKET WITH A LENGTH OF 20 AND A PITCH OF 30.

6. ADD A WELD NOTE FOR 8 MM INTERMITTENT FILLET WELDS ON THE BOTH JOINTS OF THE BRACKET WITH A LENGTH OF 10 MM AND A PITCH OF 40.

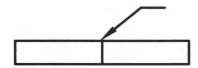

7. ADD A WELD NOTE FOR A SQUARE GROOVE WELD WITH A SEPARATION OF 6 MM ON THE TOP SURFACE OF THE PLATE.

8. ADD A WELD NOTE FOR A 6 MM V—GROOVE WELD ON THE TOP SURFACE OF THE PLATE.

OPEN THE FILE COMP89 AND DRAW THE FOLLOWING PATTERN FOR A PAPER AIRPLANE. DO
NOT INCLUDE DIMENSIONS. THE DASHED LINES REPRESENT THE FOLD LINES. NOTE THE
STARTING POINT. OBJECT IS SYMMETRICAL ABOUT LINE AB.

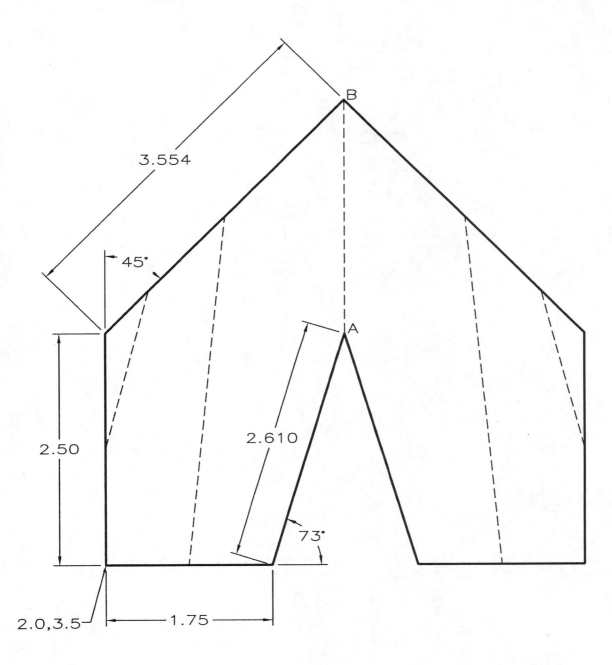

OPEN THE FILE COMP90.
DRAW THE LIGHT SWITCH PLATE, OMIT DIMENSIONS.

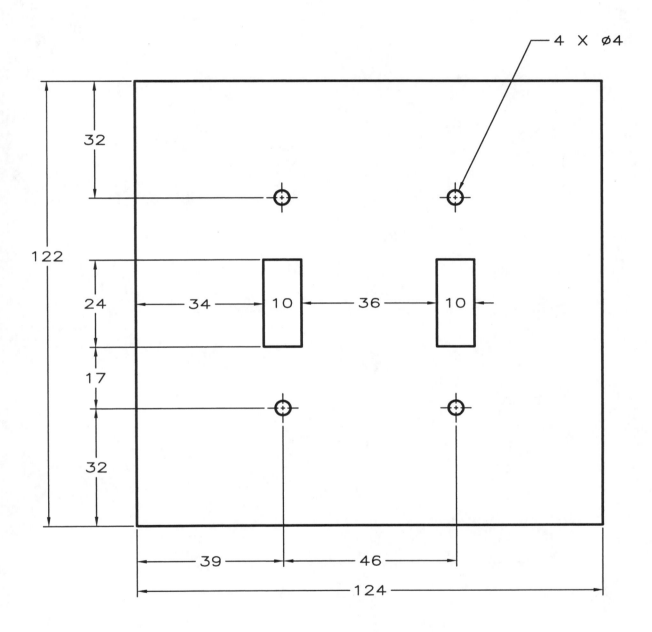

4 X Ø4

32

122

24   34   10   36   10

17

32

39   46

124

SI

| NAME: | INTRODUCTION TO | PLATE |
| SECTION: | MODERN GRAPHICS | 90 |

OPEN THE FILE COMP91 AND DRAW THE PLASTIC TAPE
DISPENSER CASE.   OMIT DIMENSIONS.

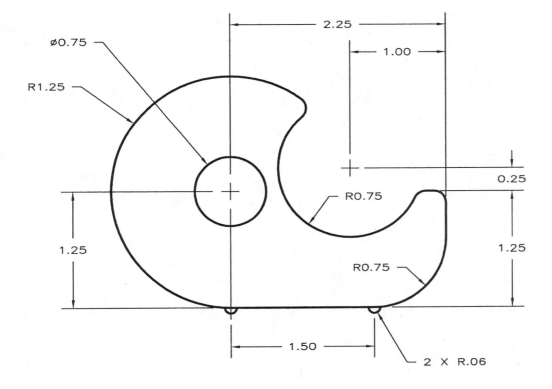

ALL ROUNDS R.125

OPEN COMP92 AND COMPLETE THE TITLE BLOCK AND PARTS LIST SHOWN
BELOW.   CREATE THREE DIFFERENT TEXT STYLES FOR EACH OF THE DIFFERENT
TEXT FONTS USED.

TEXT AT	FONT	HEIGHT	WIDTH FACTOR	OBLIQUE ANGLE	ROTATION
1	ROMAN SIMPLEX	.09	1.3	0°	0°
2	ROMAN SIMPLEX	.06	1.3	0°	0°
3	TIMES NEW ROMAN	.12	1.0	15°	0°
4	ROMAN DUPLEX	.18	1.3	15°	0°
5	ROMAN SIMPLEX	.12	1.3	0°	0°

7	CAP SCREW		1	STEEL	①
6	HANDLE CAP		1	BRASS	
5	HANDLE		1	STEEL	
4	PAD		1	RUBBER	
3	LOWER JAW		1	BRASS	
2	ADJUSTING SCREW		1	STEEL	
1	FRAME		1	CI	
NO.	PART NAME		REQ'D	MATERIAL	②

LOUISIANA STATE UNIVERSITY
BATON ROUGE, LOUISIANA   ③

DRAWING   TITLE   ④

NAME:		SEC: ⑤
2010-04-28	SCALE:	NO.:

CURRENT
DATE IN
PROPER
FORMAT

OPEN THE FILE COMP93. IN THE FILE WILL BE SEVEN PIECES TO A PUZZLE AND THE
OUTLINE FOR THE FINISHED PUZZLE. USE MODIFY COMMANDS AND OBJECT SNAP
OPTIONS TO COMPLETE THE PUZZLE. THE SOLUTION TO THE PUZZLE IS GIVEN BELOW.

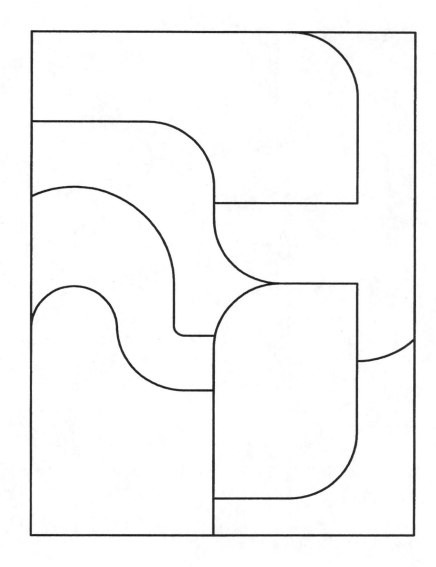

OPEN COMP94 AND DRAW THE DESIGN GIVEN.   THE CIRCLE IS GIVEN.   CREATE
LAYERS FOR EACH LINETYPE AND ASSIGN DIFFERENT COLORS AS NOTED BELOW.
    16 SIDED POLYGON AND LINES ARE RED WITH VISIBLE LINES
    OCTAGON IS BLUE WITH CENTER LINE
    SQUARE IS MAGENTA WITH DOT2 LINES
    TRANGLES ARE GREEN WITH HIDDEN LINES

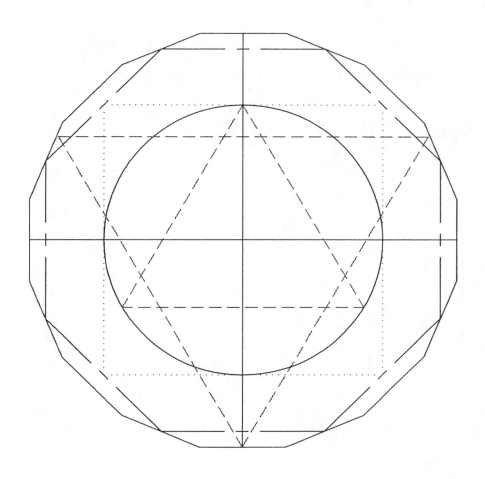

| NAME: | INTRODUCTION TO | PLATE |
| SECTION: | MODERN GRAPHICS | 94 |

OPEN THE DRAWING FILE COMP95 AND CONSTRUCT THE FOLLOWING OBJECTS.
OMIT THE DIMENSIONS.

1.

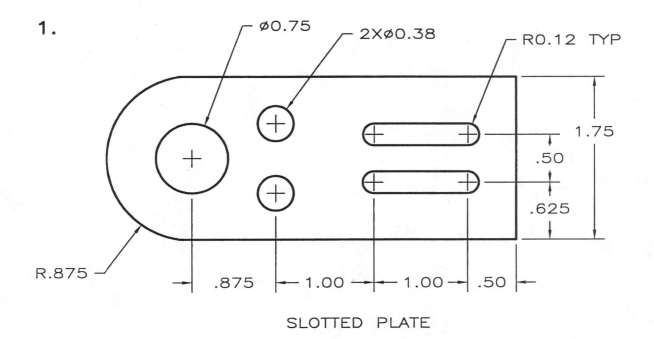

SLOTTED PLATE

2.

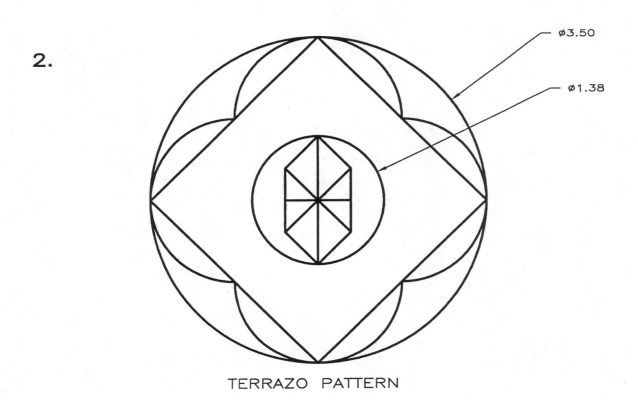

TERRAZO PATTERN

OPEN THE FILE COMP96 AND YOU WILL SEE THE FOLLOWING DRAWING.
THIS DRAWING IS NOT COMPLETE. ADD FEATURES AND MODIFY THE VIEWS AS
NEEDED SO THAT THE DRAWING DUPLICATES THE SOLUTION GIVEN BELOW.

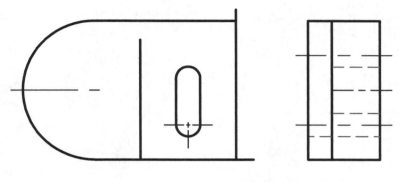

INCOMPLETE DRAWING

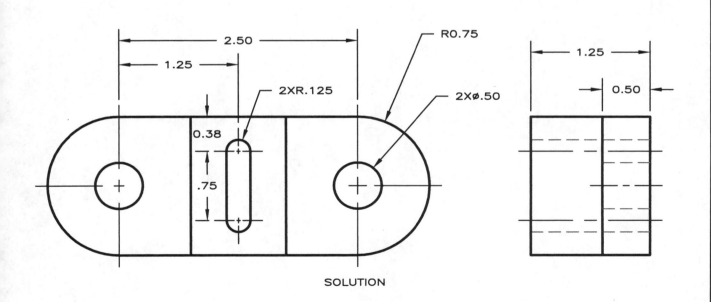

SOLUTION

OPEN THE FILE COMP97 AND COMPLETE THE PROBLEMS BELOW.

1.   COMPLETE THE DRAWING OF THE CLOCK FACE USING THE POINT STYLE SHOWN.

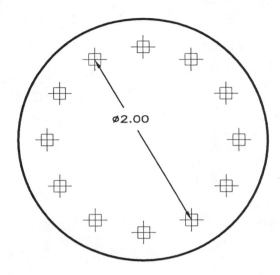

Ø2.00

2.   COMPLETE THE SOAP DISH.

.20

20

.2X.1

3.   COMPLETE THE DOOR MAT.

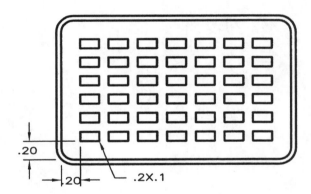

R.25

15°

OPEN THE FILE COMP98 AND CONSTRUCT THE OBJECT SHOWN BELOW.

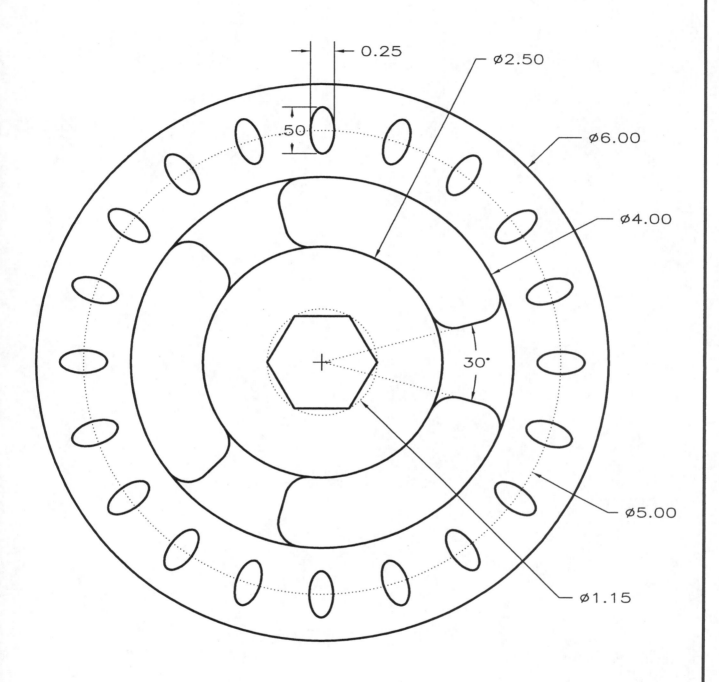

WATER VALVE
SCALE 2=1

ALL FILLETS AND ROUNDS R .25

NAME:
SECTION:

INTRODUCTION TO
MODERN GRAPHICS

PLATE
98

OPEN THE FILE COMP99.

MAKE A COPY OF THE TITLEBLOCK FROM COMP92.

1. MAKE TWO BLOCKS AND WBLOCKS CREATING "TITLE1" AND "PARTS LIST". DEFINE THE
INSERTION BASE POINTS WHERE INDICATED WITH ⊗.

2. TO CREATE "TITLE2", DELETE THE TOP PORTION OF "TITLE1" INCLUDING THE DRAWING
TITLE, SCHOOL AND THE PARTS LIST. ADD A 3"X.65" RECTANGLE TO THE LEFT SIDE OF THE
TITLE1 (SHOWN BELOW). MODIFY THE TEXT FONT FOR ALL TEXT IN TITLE 2 TO CITY
BLUEPRINT WITH A WIDTH FACTOR OF 1.3. ADD A DRAWING TITLE WITH TEXT HEIGHT OF .18.

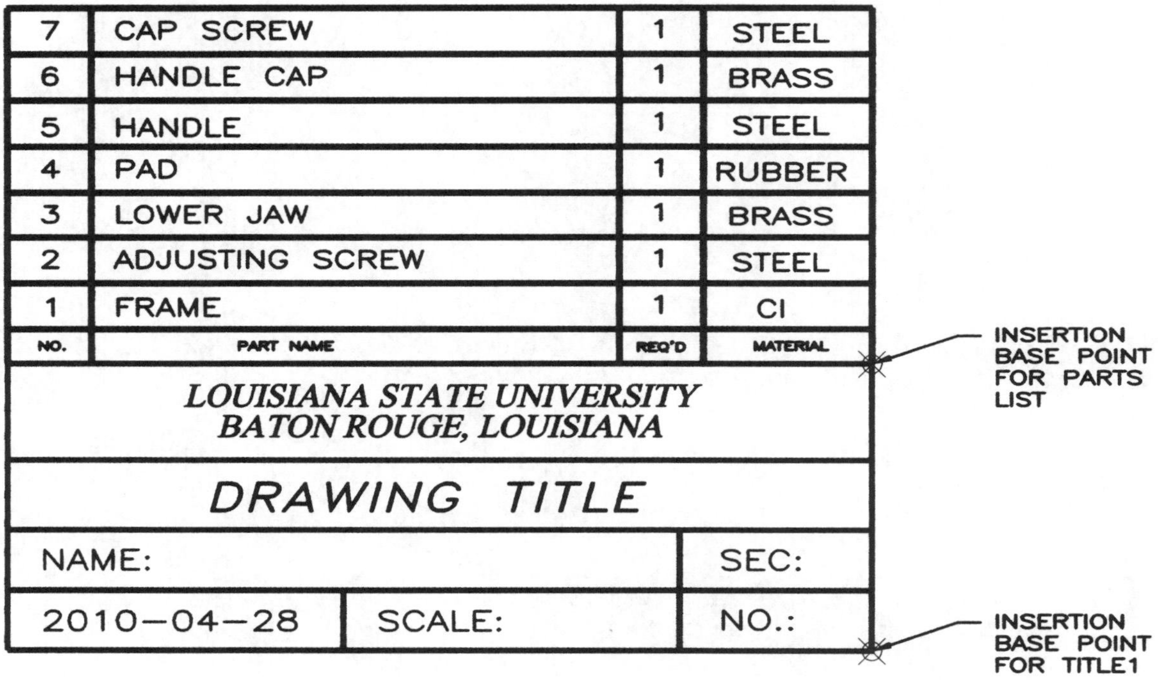

7	CAP SCREW	1	STEEL
6	HANDLE CAP	1	BRASS
5	HANDLE	1	STEEL
4	PAD	1	RUBBER
3	LOWER JAW	1	BRASS
2	ADJUSTING SCREW	1	STEEL
1	FRAME	1	CI
NO.	PART NAME	REQ'D	MATERIAL

INSERTION BASE POINT FOR PARTS LIST

LOUISIANA STATE UNIVERSITY
BATON ROUGE, LOUISIANA

DRAWING   TITLE

NAME:		SEC:
2010—04—28	SCALE:	NO.:

INSERTION BASE POINT FOR TITLE1

TITLE1

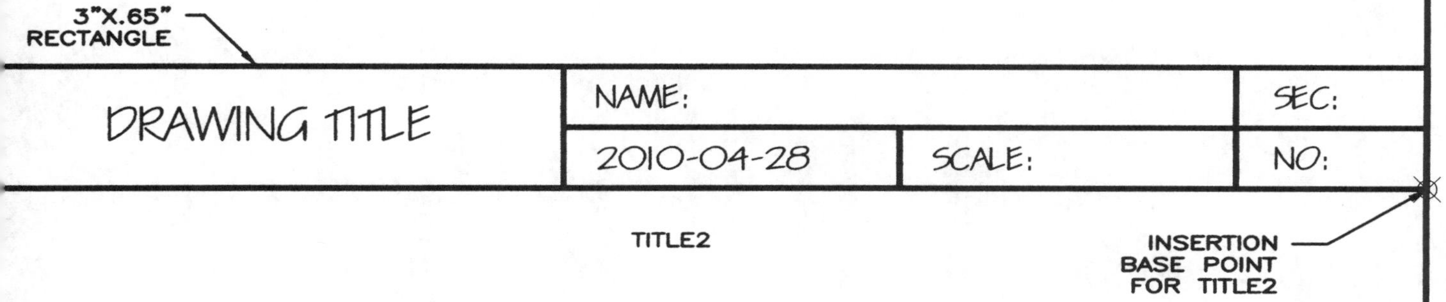

3"X.65" RECTANGLE

DRAWING TITLE	NAME:		SEC:
	2010-04-28	SCALE:	NO:

TITLE2

INSERTION BASE POINT FOR TITLE2

NAME:
SECTION:

INTRODUCTION TO
MODERN GRAPHICS

PLATE
99

USE THE TWO TITLE BLOCKS CREATED ON PLATE 99 TO CREATE THE 4 SIZE A SHEETS SHOWN.
THE RECTANGLES ARE 7.5X10 OR 10X7.5, ALLOWING A HALF INCH BORDER AROUND THE PAGE.
CREATE A SEPARATE FILE FOR EACH PROBLEM AND NAME THE FILE AS GIVEN BELOW.

NOTE: 1.   VERTICAL ORIENTATION — INSERT TITLE1 AND REDUCE IT 25%.
        2.   HORIZONTAL ORIENTATION — INSERT TITLE1 FULL SIZE.
        3.   VERTICAL ORIENTATION — INSERT TITLE2 FULL SIZE.
        4.   HORIZONTAL ORIENTATION — INSERT TITLE 2 AND INCREASE IT 33%.

LOUISIANA STATE UNIVERSITY
BATON ROUGE, LOUISIANA
DRAWING TITLE

| NAME: | | SEC: |
| 2010—04—28 | SCALE: | NO.: |

1.   8X11—T1

LOUISIANA STATE UNIVERSITY
BATON ROUGE, LOUISIANA
DRAWING TITLE

| NAME: | | SEC: |
| 2010—04—28 | SCALE: | NO.: |

2.   11X8.5—T1

| DRAWING TITLE | NAME: | | SEC: |
| | 2010-04-28 | SCALE: | NO.: |

3.   8X11—T2

| DRAWING TITLE | NAME: | | SEC: |
| | 2010-04-28 | SCALE: | NO.: |

4.   11X8.5—T2

NAME:
SECTION:

INTRODUCTION TO
MODERN GRAPHICS

PLATE
100

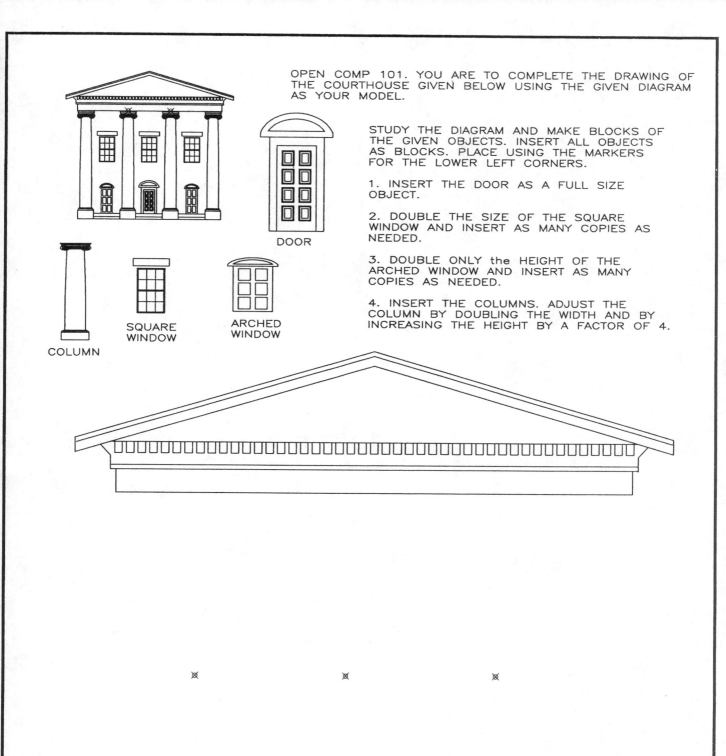

OPEN COMP 101. YOU ARE TO COMPLETE THE DRAWING OF THE COURTHOUSE GIVEN BELOW USING THE GIVEN DIAGRAM AS YOUR MODEL.

STUDY THE DIAGRAM AND MAKE BLOCKS OF THE GIVEN OBJECTS. INSERT ALL OBJECTS AS BLOCKS. PLACE USING THE MARKERS FOR THE LOWER LEFT CORNERS.

1. INSERT THE DOOR AS A FULL SIZE OBJECT.

2. DOUBLE THE SIZE OF THE SQUARE WINDOW AND INSERT AS MANY COPIES AS NEEDED.

3. DOUBLE ONLY the HEIGHT OF THE ARCHED WINDOW AND INSERT AS MANY COPIES AS NEEDED.

4. INSERT THE COLUMNS. ADJUST THE COLUMN BY DOUBLING THE WIDTH AND BY INCREASING THE HEIGHT BY A FACTOR OF 4.

DOOR

SQUARE WINDOW

ARCHED WINDOW

COLUMN

NAME:
SECTION:

INTRODUCTION TO MODERN GRAPHICS

PLATE 101

OPEN THE FILE COMP102.

COMPLETE PROBLEM 1 AS A FULL SECTION AND PROBLEM 2 AS A HALF SECTION.

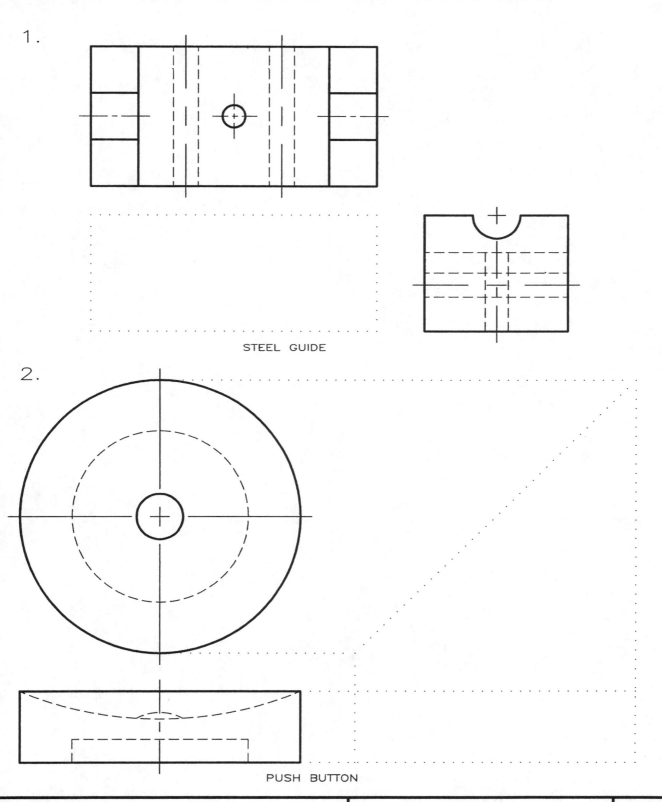

1.

STEEL GUIDE

2.

PUSH BUTTON

NAME:
SECTION:

INTRODUCTION TO
MODERN GRAPHICS

PLATE
102

OPEN THE FILE COMP103.

CONSTRUCT THE FRONT VIEW OF THE KNOB AS A HALF SECTION.

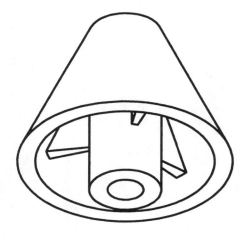

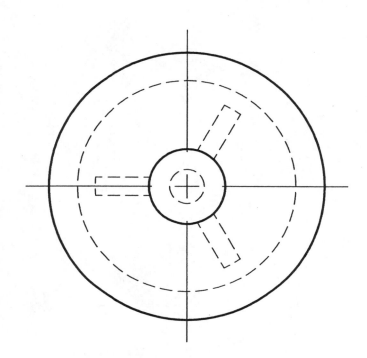

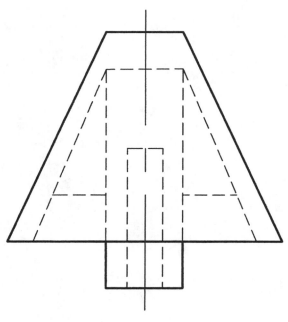

KNOB
GLASS
SCALE 2:1

NAME:
SECTION:

INTRODUCTION TO
MODERN GRAPHICS

PLATE
103

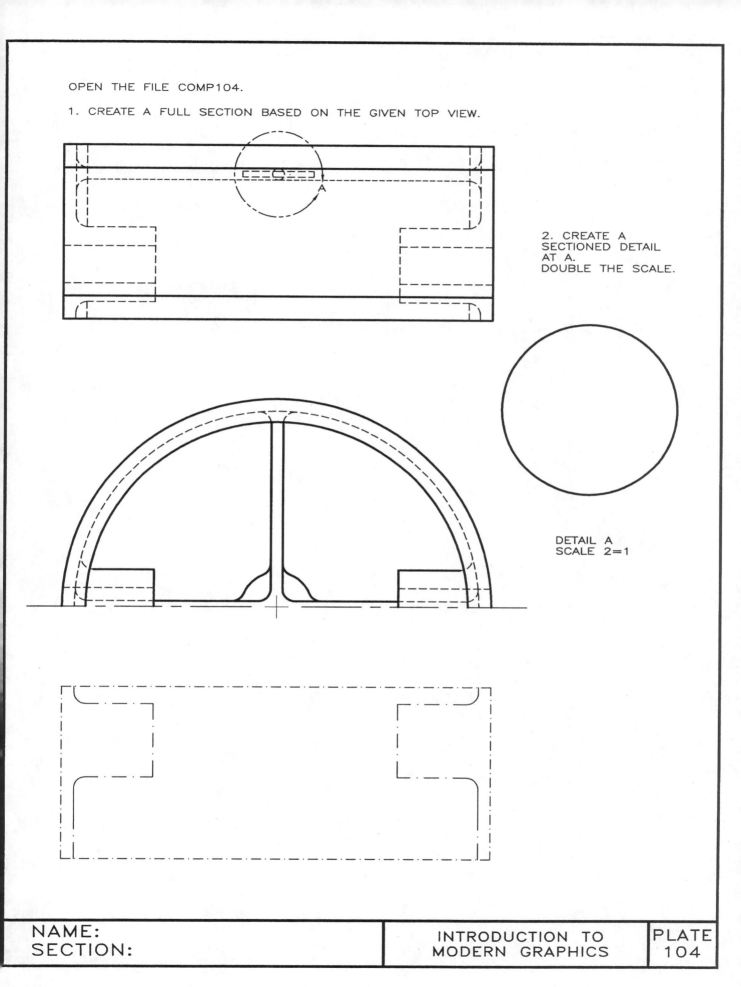

OPEN THE FILE COMP104.

1. CREATE A FULL SECTION BASED ON THE GIVEN TOP VIEW.

2. CREATE A SECTIONED DETAIL AT A. DOUBLE THE SCALE.

A

DETAIL A
SCALE 2=1

NAME:
SECTION:

INTRODUCTION TO
MODERN GRAPHICS

PLATE
104

OPEN THE FILE COMP 105.
1. DRAW THE INDICATED REVOLVED AND REMOVED SECTIONS FOR THE
WATCH STEM GIVEN BELOW.

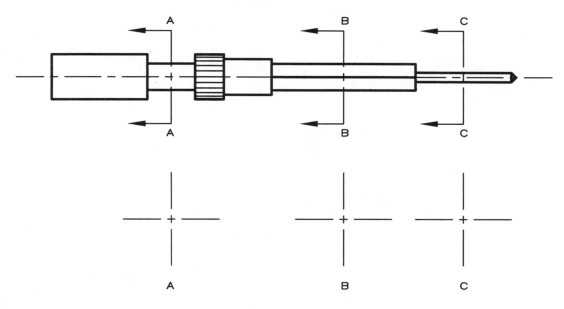

2. DRAW THE INDICATED REMOVED SECTIONS FOR THE JEWELER'S
SCREW DRIVER GIVEN BELOW.

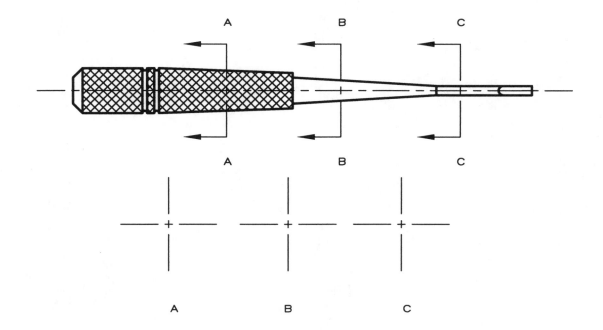

NAME:
SECTION:

INTRODUCTION TO
MODERN GRAPHICS

PLATE
105

OPEN THE FILE COMP106.
MAKE AN ASSEMBLY OF THE C CLAMP. INCLUDE PART LABELS AND PARTS LIST.

6 HANDLE CAP
STEEL
2 REQ.

7 CAP SCREW
STEEL
1 REQ.

3 LOWER JAW
STEEL
1 REQ.

4 PAD
RUBBER
2 REQ.

5 HANDLE
STEEL
1 REQ.

1 FRAME
CAST IRON
1 REQ.

2 ADJUSTING SCREW
STEEL
1 REQ.

NAME:
SECTION:

INTRODUCTION TO
MODERN GRAPHICS

PLATE
106

OPEN THE FILE COMP107.
DIMENSION THE FOLLOWING THE OBJECTS.

① CONE

② SUPPORT

③ CYLINDER

NAME:
SECTION:

INTRODUCTION TO
MODERN GRAPHICS

PLATE
107

OPEN THE FILE COMP108.

COMPLETELY DIMENSION THE OBJECT BELOW.

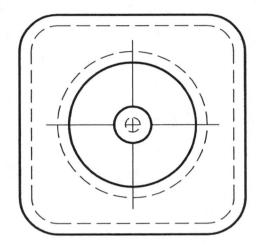

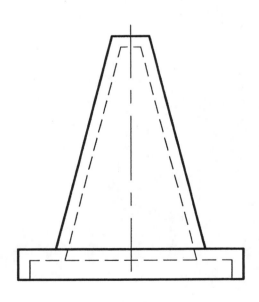

SAFETY CONE
PLASTIC
SCALE: 1:10

SI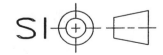

NAME:

SECTION:

INTRODUCTION TO
MODERN GRAPHICS

PLATE
108

OPEN THE FILE COMP109.

COMPLETELY DIMENSION THE OBJECT SHOWN BELOW.

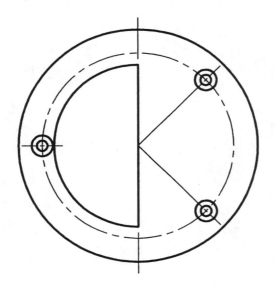

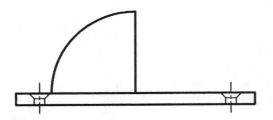

DOOR STOP
STEEL

NAME:
SECTION:

INTRODUCTION TO
MODERN GRAPHICS

PLATE
109

OPEN THE FILE COMP110.

DIMENSION THE BRICK USING THE SCALE GIVEN.

FROGGED BRICK
1"=2"

NAME:
SECTION:

INTRODUCTION TO
MODERN GRAPHICS

PLATE
110

OPEN THE FILE COMP111.

DIMENSION PART 1 AND COMPLETE A FULL SECTION OF THE FRONT VIEW.
CONSTRUCT SCHEMATIC THREADS FOR PART 2.

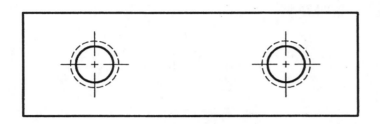

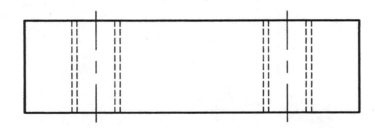

(1) BLOCK
       STEEL
       1 REQ'D

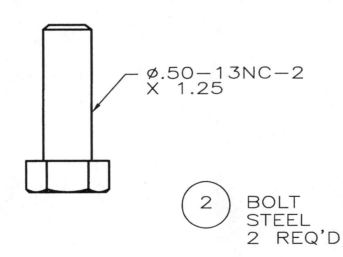

Ø.50−13NC−2
X 1.25

(2) BOLT
       STEEL
       2 REQ'D

NAME:
SECTION:

INTRODUCTION TO
MODERN GRAPHICS

PLATE
111

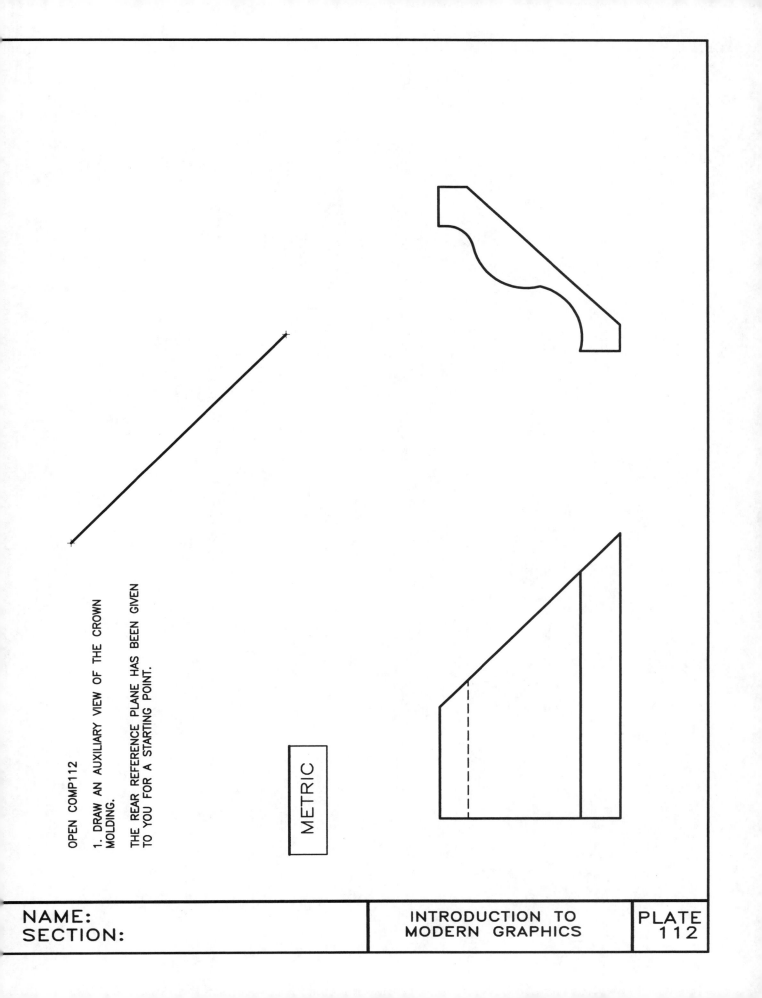

OPEN COMP112

1. DRAW AN AUXILIARY VIEW OF THE CROWN MOLDING.

THE REAR REFERENCE PLANE HAS BEEN GIVEN TO YOU FOR A STARTING POINT.

METRIC

NAME:
SECTION:

INTRODUCTION TO
MODERN GRAPHICS

PLATE
112

CONSTRUCT THE BASKETBALL HALF COURT BELOW.
PLOT SCALE ⅛"=1'−0.

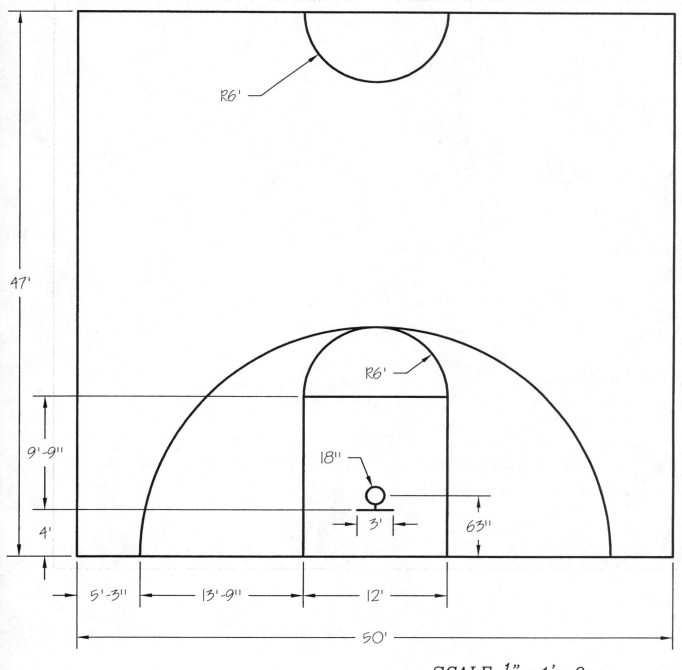

R6'

R6'

18"

47'

9'-9"

4'

3'

63"

5'-3"

13'-9"

12'

50'

SCALE ⅛*" =1' −0*

USE THE DRAWING FROM PLATE 113 TO CREATE THE LAYOUT FOR TWO FULL COURTS AS
SHOWN BELOW. ADD THE SCORERS' TABLES AND RESTRAINING LINES. SCALE $\frac{1}{16}$"=1'-0.

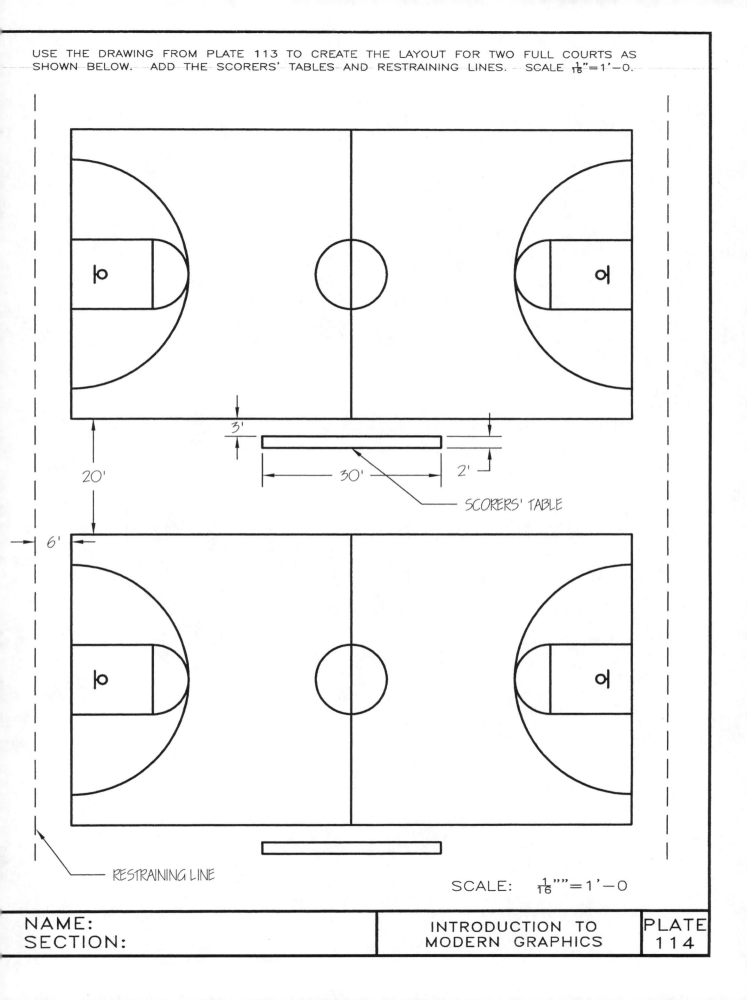

SCORERS' TABLE

RESTRAINING LINE

SCALE: $\frac{1}{16}$""=1'-0

NAME:
SECTION:

INTRODUCTION TO
MODERN GRAPHICS

PLATE
114

CONSTRUCT SOLIDMODELS FOR EACH OF THE PROBLEMS.   GRID = .25" OR 6MM.

1.

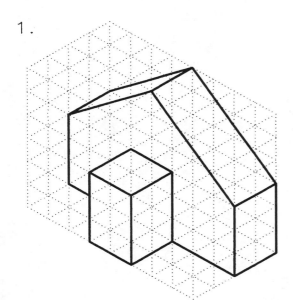

2.

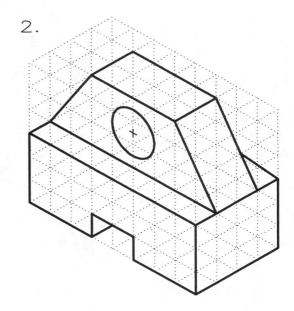

3.

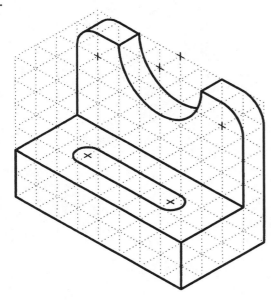

4.

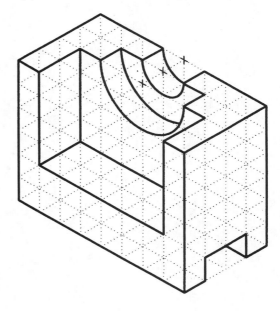

NAME:
SECTION:

INTRODUCTION TO
MODERN GRAPHICS

PLATE
115

CREATE A SOLID MODEL FOR EACH OF THE FOLLOWING OBJECTS. ASSUME ALL HOLES
AND SLOTS EXTEND THROUGH THE OBJECTS. GRID = .25 OR 6MM.

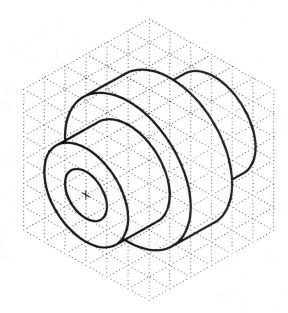

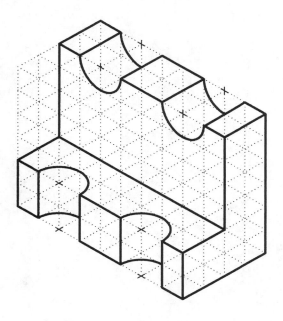

(1) PULLEY

(2) BRACE

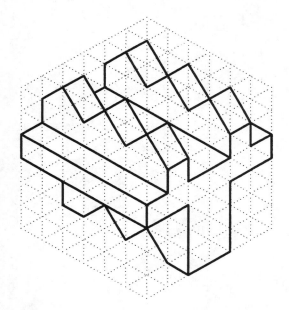

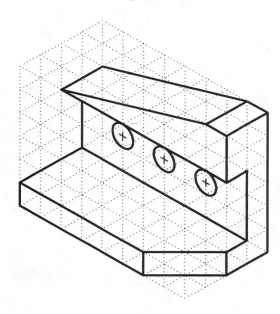

(3) PACKING FORM

(4) SLIDE

NAME:
SECTION:

INTRODUCTION TO
MODERN GRAPHICS

PLATE
116

CREATE A SOLID MODEL OF THE MANTLE SHOWN BELOW.   ADD DENTAL MOLDING AND
OTHER DECORATIVE FEATURES.

NOTE: GRID = 6".

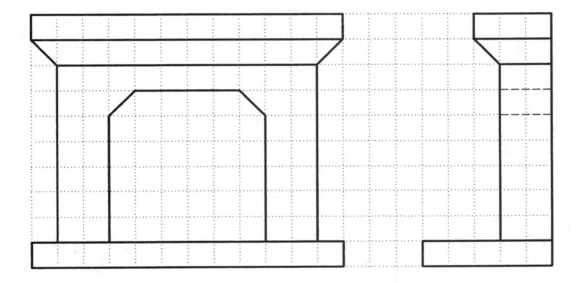

NAME: ....

SECTION: ....

INTRODUCTION TO
MODERN GRAPHICS

PLATE
117

CONSTRUCT A SOLID MODEL OF THE MOUNTING
PLATE.   CREATE AN ENGINEERING DRAWING OF
THE MODEL.   BE SURE TO INCLUDE
DIMENSIONS.

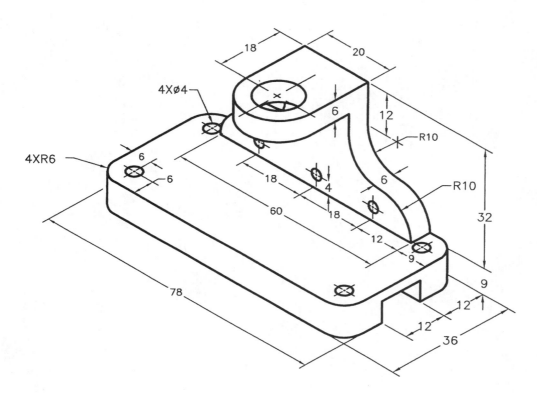

SI

NAME: ....
SECTION: ....

INTRODUCTION TO
MODERN GRAPHICS

PLATE
118

OPEN THE FILE COMP 119.   USING THE GIVEN INFORMATION COMPLETE THE
DRAWING OF THE SPUR GEAR.   CALCULATE THE NEEDED VALUES.

NUMBER OF TEETH = 20
PITCH DIAMETER = 6:00
DIAMETRAL PITCH = 5.00
PRESSURE ANGLE = 20°
r=.312
R=.804

1. ADDENDUM =

2. DEDENDUM =

3. OUTSIDE DIAMETER =

4. ROOT DIAMETER =

5. WHOLE DEPTH =

6. CHORDAL THICKNESS =

7. CHORDAL ADDENDUM =

8. WORKING DEPTH =

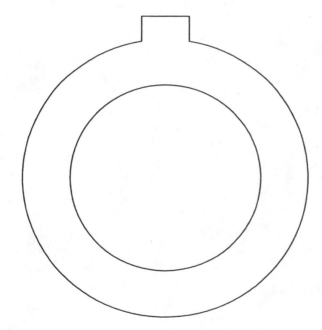

NAME:
SECTION:

INTRODUCTION TO
MODERN GRAPHICS

PLATE
119

THIS PROBLEM WILL REVIEW THE ACAD COMMANDS THAT YOU HAVE MASTERED. YOU WILL USE YOUR SKILLS TO CREATE A DRAWING OF THE WHITE HOUSE. FOLLOW THE DIRECTIONS AND YOU WILL BE ABLE TO COMPLETE THIS ASSIGNMENT QUICKLY.

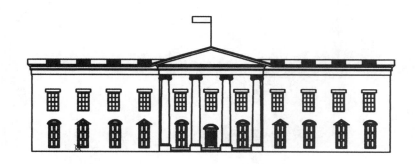

1. OPEN THE FILE COMP 120. YOU WILL SEE A BLANK PAGE.

2. YOU WILL INSERT THE BLOCK "COURTHOUSE" INTO THIS FILE AND YOU NEED TO EXPLODE IT. PLACE THIS BLOCK OUTSIDE OF YOUR DRAWING SHEET.

| NAME: | INTRODUCTION TO | PLATE |
| SECTION: | MODERN GRAPHICS | 120 |

3. MAKE A COPY OF THE COURTHOUSE BLOCK. YOU WILL NEED THIS LATER IN THE PROJECT.

4. ON ONLY ONE COPY OF THE COURTHOUSE BLOCK ERASE THE COLUMNS. LEAVE THE SMALL DETAILS AT THE EDGE OF THE ROOF.

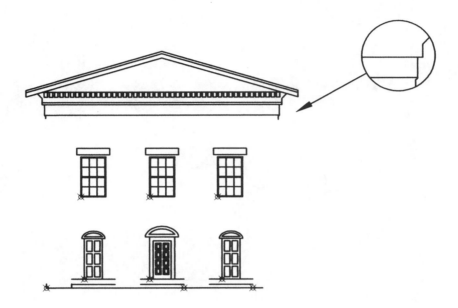

5. REMOVE THE ROOF ABOVE THE MOLDING

6. REMOVE THE FRONT DOOR BUT LEAVE THE BASE AND THE NODE POINTS

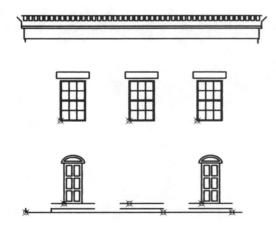

NAME:
SECTION:

INTRODUCTION TO
MODERN GRAPHICS

PLATE
121

7. INSERT THE BLOCK "TRIANGLE WINDOW" AT THE OLD NODE POINT FOR THE DOOR.

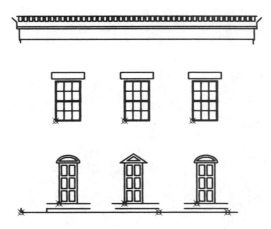

8. COPY THE CENTER COLUMN OF WINDOWS.   USE THE NODE POINT FOR THE FIRST COLUMN AS THE FIRST DISPLACEMENT POINT AND THE NODE POINT AT THE WINDOW AS THE SECOND DISPLACEMENT POINT.

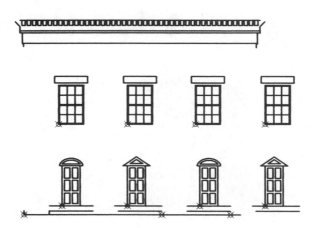

9.   MARK THE MIDPOINT OF THE ROOF WITH A SINGLE POINT AND DRAW A VERTICAL LINE THROUGH THE ENTIRE DRAWING AT THIS POINT.

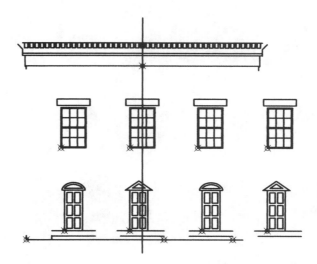

NAME:
SECTION:

INTRODUCTION TO
MODERN GRAPHICS

PLATE
122

10. STRETCH THE HORIZONTAL LINES AT THE ROOF 1.6 INCHES.   USE THE MIDPOINT AS
THE BASE POINT

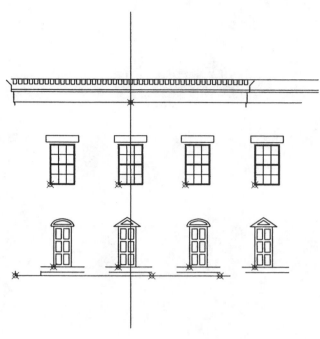

11. MOVE THE EDGES OF THE ROOF TO THE NEW END POINTS

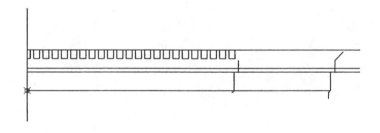

12. ERASE ANY STRAY LINES

NEXT YOU WILL WORK TO COMPLETE THE MOLDING AT THE ROOF LINE. YOU MAY WANT TO
ERASE THE TOP LINE OF THIS MOLD AND DRAW IT BACK AS A SINGLE LINE WHEN YOU ARE
FINISHED. (YOU ONLY NEED THIS STEP IF YOUR PRINTER IS PRINTING THIS AS DOUBLE
LINES)

NAME:
SECTION:

INTRODUCTION TO
MODERN GRAPHICS

PLATE
123

13. ARRAY THE MOLDING TO COMPLETE THE ROOF LINE. PICK OFFSET POINTS ON THE SCREEN. YOU WILL NEED 11 COLUMNS AND 1 ROW. WHEN YOU ARE FINISHED, ERASE THE MIDPOINT LINE IF YOU HAVE NOT ALREADY DONE SO.

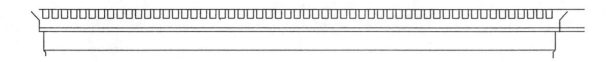

14. EXTEND AND TRIM LINES AT THE TOP EDGES OF THE ROOF TO COMPLETE THIS SECTION.

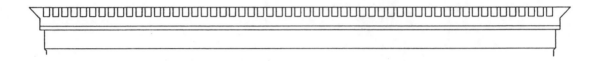

15. EXPLODE BLOCK WINDOWS ON FIRST STORY AND ERASE STEP LINES AND NODE POINTS AT WINDOWS.

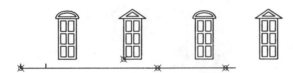

16. ERASE STRAY LINES AND NODE POINTS AND VERTICAL LINES FOR WALL EDGES. YOU MAY WANT TO ERASE AND REDRAW THE BASE SO THAT IT WILL BE ONE LONG SEGMENT.

NAME:
SECTION:

INTRODUCTION TO
MODERN GRAPHICS

PLATE
124

17. DRAW VERTICAL LINES AT THE MIDPOINT OF EACH WINDOW ON THE TOP ROW. INSERT THE BLOCK "ROOF RAIL" AT EACH OF THESE MIDPOINTS.

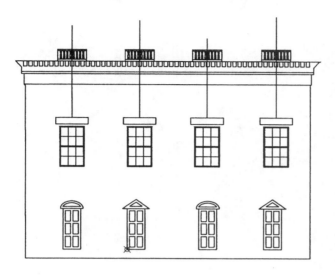

18. DRAW EDGES BETWEEN THE RAIL BLOCKS.

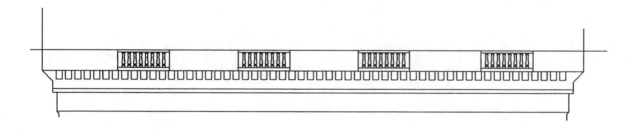

19. ERASE ANY STRAY MARKS. YOUR DRAWING SHOULD LOOK LIKE THIS.

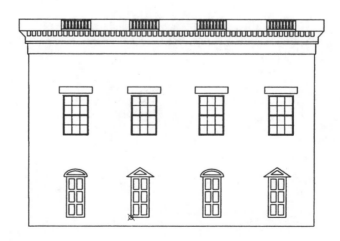

NAME:
SECTION:

INTRODUCTION TO
MODERN GRAPHICS

PLATE
125

20. PLACE YOUR DRAWING NEXT TO THE COURTHOUSE BLOCK THAT YOU DID NOT USE. USE THE LOWER ENDPOINTS TO LINE UP THE BOTTOM EDGE.

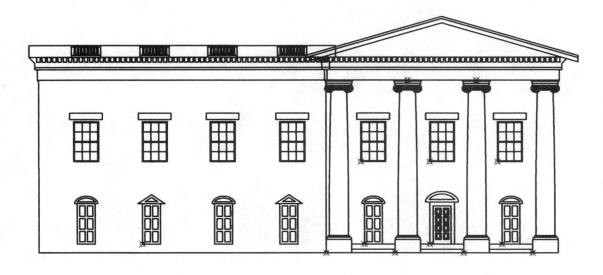

21. ERASE FEATURES THAT FALL BEHIND THE ROOF.

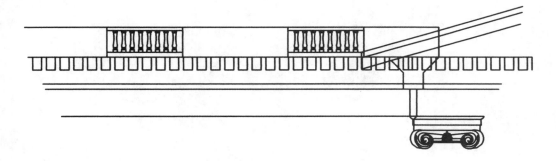

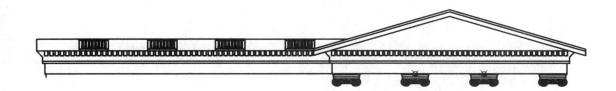

NAME:
SECTION:

INTRODUCTION TO
MODERN GRAPHICS

PLATE
126

22. SCALE ALL THE WINDOWS ON THE FIRST FLOOR BY 1.3 USING THE LOWER LEFT CORNER OF EACH WINDOW AS YOUR BASE POINT. (DO THESE ONE AT TIME TO MAINTAIN THE SPACING.

23. MIRROR THE WING USING THE A STRAIGHT VERTICAL LINE THROUGH THE PEAK OF THE ROOF FOR YOUR MIRROR LINE

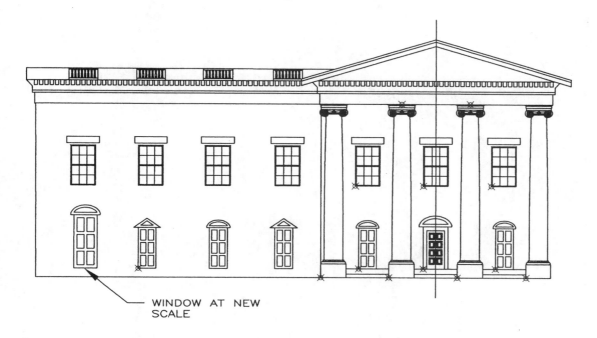

WINDOW AT NEW SCALE

24. YOUR DRAWING NOW LOOKS LIKE THIS.

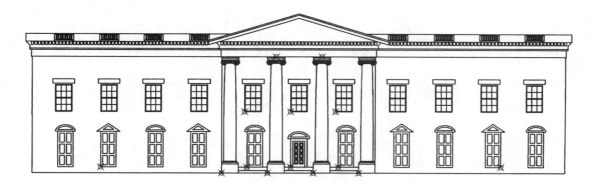

25. YOU MY WANT TO ADD A FLAG AT THE TOP.

NAME:
SECTION:

INTRODUCTION TO
MODERN GRAPHICS

PLATE
127

26. SCALE YOUR DRAWING BY 45 PERCENT AND ROTATE IT 90 DEGREES.

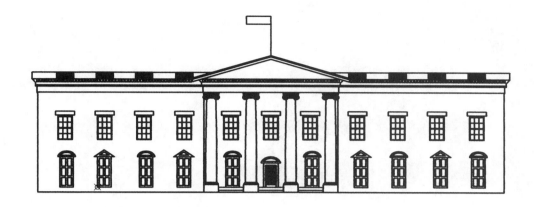

27. YOUR DRAWING OF THE WHITE HOUSE IS NOW FINISHED AND IT LOOKS LIKE THIS:

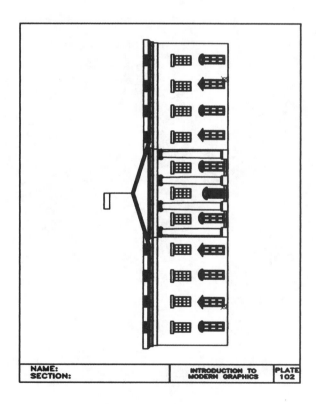

NAME:
SECTION:

INTRODUCTION TO
MODERN GRAPHICS

PLATE
128

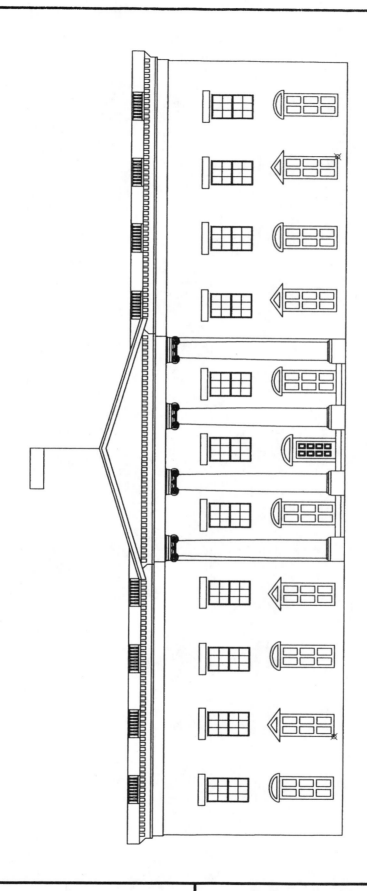

NAME:
SECTION:

INTRODUCTION TO
MODERN GRAPHICS

PLATE
129

NAME: ....
SECTION: ....

INTRODUCTION TO
MODERN GRAPHICS

PLATE

INTRODUCTION TO
MODERN GRAPHICS

PLATE

INTRODUCTION TO
MODERN GRAPHICS

INTRODUCTION TO
MODERN GRAPHICS

INTRODUCTION TO
MODERN GRAPHICS

PLATE

NAME: ....

SECTION: ....

INTRODUCTION TO
MODERN GRAPHICS

PLATE

NAME:

SECTION:

INTRODUCTION TO
MODERN GRAPHICS

PLATE